KU-259-866

GREAT MEAT

CLASSIC **TECHNIQUES** AND AWARD-WINNING **RECIPES** FOR SELECTING, CUTTING, AND COOKING **BEEF**, **LAMB**, **PORK**, **POULTRY,** AND **GAME**

DAVE KELLY OF **RUBY & WHITE**

Foreword and contributed recipes by
GLENN KEEFER, Owner, and
JOHN HOGAN, Executive Chef,
KEEFER'S of Chicago

FAIR WINDS
P R E S S
BEVERLY, MASSACHUSETTS

First published in the USA in 2013
by Fair Winds Press, a member of
Quayside Publishing Group
100 Cummings Center
Suite 406-L
Beverly, MA 01915-6101
www.fairwindspress.com

© 2013 Fair Winds Press, an imprint of
Rockport Publishers, Inc. and Toucan Books
Ltd. All rights reserved. No part of this book
may be reproduced or utilized, in any form
or by any means, electronic or mechanical,
without prior permission in writing from
the publisher.

17 16 15 14 13 1 2 3 4 5

ISBN: 978-1-59233-581-7

Digital edition published in 2013
eISBN: 978-1-61058-884-3

For Toucan Books

Managing Editor	Ellen Dupont
Editor	Judith Samuelson
Designer	Lee Riches
Special photography	Mark Winwood
Proofreader and indexer	Andrea Chesman
Additional writing	Genevieve Taylor
Additional recipes	Jacqueline Bellefontaine

Printed and bound in China

GREAT
MEAT

Leabharlann Shráid Chaoimhín
Kevin Street Library
Tel: 01 222 8488

Contents

Foreword

The art of selecting great meat means selecting a purveyor you can trust to deliver consistent quality day in and day out. I've always chosen to work with artisans and suppliers who are as dedicated to their craft as we are to ours. I've spent my entire career searching for and selecting the best meat available. When I was asked to participate in the creation of this book I felt an immediate kinship with the artisans at Ruby & White, despite our living on opposite sides of The Big Drink.

Our quest to find and serve great meat began the same way—by establishing unwavering standards, then finding like-minded farmers and ranchers who could meet or exceed those standards with consistency and integrity.

My own journey to find great meat began twenty-one years ago, when I was the general manager of the Chicago Ruth's Chris Steakhouse. I invited the three top suppliers of beef to bring their best steaks. These were companies who, at that time, supplied the three best steakhouses in Chicago: Morton's, Ruth's, and Gibsons. It was a blind taste test—only our broiler cook knew which steak was which. The owners of Ruth's and I tasted the steaks, and we all chose the same steak as the best. I still buy from this same supplier 21 years later. The meat purveyor-restaurant relationship is unique. It assumes trust but is built on a set of control criteria that makes a prenuptial agreement look friendly.

Keefer's welcomes diners who appreciate the unrivalled quality of its meat.

PRIMED FOR SUCCESS
Great steak has lots of marbling and is graded as USDA prime. Only two percent of the beef produced in the United States earns the grade of prime. Steaks should have generous fat content, not in the form of

translucent, grisly connective tissue, but rather lots of flecks of snow-white fat throughout the meat. The fat should melt to the touch with the heat of your hand. Enzymes in this lush meat break down the fibers during the aging process.

There are two forms of aging: wet or dry. How the meat is aged makes a big difference in the flavor. Ideally, dry-aging is done in a walk-in cooler lined with cedar or redwood with no moisture. Some come outfitted with black lights rather than incandescent bulbs to eliminate condensation and the opportunity for mold. The meat, still in primal cuts (large, distinct sections), is hung so as to get maximum ventilation; some use fans. Dry-aging makes the meat shrink but yields a concentrated, robust flavor that's tangy and rich. Wet-aged beef is aged enclosed in plastic and has a sweeter flavor profile with more juices. In both cases, aging should take place at very low temperatures.

Both Keefer's and Ruby & White age our beef for approximately 30 days. There are some who age upwards of 50 days, which can be downright offensive if you are not accustomed to heavily aged beef, though some love that funky flavor.

STEAK NEEDS A LITTLE TLC

The trick to serving great steak is in the handling of the meat as much as the provenance. Great steak is about the way the meat is seasoned, the temperature at which the steak is aged, the temperature at which it is cooked, how it is rested, and the temperature of the plate on which it is served, as well as where it comes from. Keefer's Executive Chef, John Hogan, uses a liberal dose of sea salt and crushed black pepper just before putting steaks on our specially built broiler, which can reach insane temperatures of 1,800 degrees Fahrenheit. Once the steak is cooked near to its desired temperature, the steak must be rested, meaning it needs to sit away from a high heat for a few minutes while the juices settle down and the interior color of the steak evens out. The center color should be evenly matched with the entire inside of the steak. Chef John then tops all steaks with Maitre d' Hotel compound butter, made with sweet butter, lemon juice, shallots, and fresh parsley.

PROVENANCE AND ORIGIN

We also buy specially raised beef from Bob and Penny Lafin of Windy Hill Farms in Grant Park, Illinois. This beef is antibiotic- and hormone-free. Bob and Penny use non-GMO corn and molasses to feed their Black Angus cattle. They are careful to use whole fresh field corn, not the corn by-product of ethanol production that some farmers use.

WELL-DESERVED PRAISE

Over the past 11 years, Keefer's Restaurant has earned the following accolades: named "Best of Chicago for Steak" by the Food Network; listed as "One of America's Top 10 Steakhouses" by *Playboy* magazine; profiled in "It's not Hip; It's Perfection" in Crain's Chicago Business, Check Please!; voted "Best Place for Business Dining" by readers of Crain's Chicago Business; named one of the restaurants of the decade by Phil Vettel of the *Chicago Tribune*; and recommended by The Michelin Guide.

Keefer's Restaurant have served both politicians and celebrities, ranging from Barack Obama to Beyoncé, but our true following comes from local Chicago business people and devoted regulars from city neighborhoods and the surrounding suburbs.

In over 30 years of serving steaks at The Palm, Ruth's Chris Steak House and Keefer's, I have witnessed the rise and fall and the rise again of the glamour of steak. People have never stopped eating well-crafted beef, even when it was considered taboo during the 1980s, when cholesterol seemed to be on people's radar for the first time. Fad diets come and fad diets go, and the number of vegans and vegetarians continues to increase, but steak has never truly fallen from favor. There is something truly American and an essence connected to success and celebration that seems to endorse steak as the meal of choice. Look at how many new steak restaurants are opening up in Chicago alone!

GLOBAL LEADER

Nowhere in the world will you get a steak like we serve here in America. Forget about that grass-fed stuff from Argentina—it lacks the fat content to stand up to the cooking process. Try one of those at any temperature over medium-rare and you will get a livery, grainy disappointment of a steak.

For me, the ultimate meal is Chef John's Celery Root Purée with Rabbit Rillettes, Brussels Sprout Salad with Burrata, Rare New York Strip Steak Au Poivre, Mushroom Potato Gratin, and a great big Malbec.

Bon Appetit!

GLENN KEEFER
Owner and General Manager, Keefer's Restaurant

A luxurious, open-plan interior and faultlessly attentive service marks out Keefer's as one of the best steakhouse experiences.

Introduction

At Ruby & White, our aim is to bring all our customers the best meat possible by using great suppliers who share our values. But we are more than just butchers; we want our customers to get the most out of the meat when they get it home. We willingly give expert advice on particular cuts of meat and explain how to cook the more uncommon, but equally delicious, cuts.

I'm hoping that this book will encourage you to try new meats, new cuts, and new recipes.

A butcher's counter should tempt and entice you.

The meat at Ruby & White is dry-aged in ideal conditions—the walk-in fridge is kept at a chilly 35–39°F (2–4°C).

PROVENANCE

My customers love hearing about the family farmers who supply us with meat and the dedicated foragers who bring in the game. After extensive searching, we have found farmers who are as passionate about meat as we are, and every link in our food chain takes pride in what we are producing. All our meat is free-range and comes from local farms. Ruby Reds and British Whites are the two beef breeds from which we took our name, but we use other breeds, too, including Hereford, South Devons, and Aberdeen Angus. These native breeds do not provide the same yield as larger cattle, but their marbling, texture, and flavor is unrivalled. Our pork comes from a carefully selected farm that uses traditional breeding and rearing methods. Raised in a natural environment and not forced to gain weight, these animals produce really delicious meat, with the perfect amount of fat.

IN THE SHOP

Sourcing great meat is only the beginning of the story. It's our job to take great meat and make it the best meat. Bar none. That's why we dry-age our beef for 28–35 days in a spacious, temperature-controlled fridge. As the meat ages, enzymes make it tender and flavorful as the carcass loses moisture. In simple terms, this works in the same way as reducing a jus to enhance a sauce. With the optimum temperature, air flow, and humidity, the meat slowly matures, becoming the best it can be. This process is time- and space-consuming, and comes at a cost, which is why so few butchers age their meat in this traditional way.

CHOOSING A BUTCHER

When you walk into a butcher's take a sniff: It will smell of meat, with no off odors. The butcher should be friendly and accommodating, someone who will bone-out a chicken or supply bones for stock. But butchery is more than just a trade—it's a skill, a craft, and requires real passion. For an independent butcher, not governed by the relentless pursuit of the bottom line, the butcher's counter is the heart and soul of the operation. The counter should be full of great stuff, possibly some unfamiliar items, and should make you want to get cooking. Look for cuts on the bone, especially beef. Whole carcasses should be on-site, so any cut should be available; and if not, should be ordered in on request. Once you've got a great butcher, you're guaranteed great meat.

Our game birds are hung undrawn to allow the flavor and texture to mature. The feathers protect the skin against drying.

ON THE PLATE

With meat, the expression "you are what you eat" couldn't be more true. As an example, our free-range lambs graze on a grass–chicory mix, which the farmer supplements with turnips whose sweetness you can actually taste in the meat. But great meat is about more than just the meat—how you cook it is vital. I've included my own favorite recipes in this book, but most are from the chefs that I supply—thanks, guys! I'm grateful to them and to my friends at Keefer's for sharing their knowledge and passion for great meat.

Enjoy!

DAVE KELLY
Ruby & White

chapter 1
Beef

Perfectly cooked and correctly aged beef—whether roasted, broiled, pot-roasted, panfried, or braised—is perhaps the pinnacle of good eating for the passionate, dedicated carnivore.

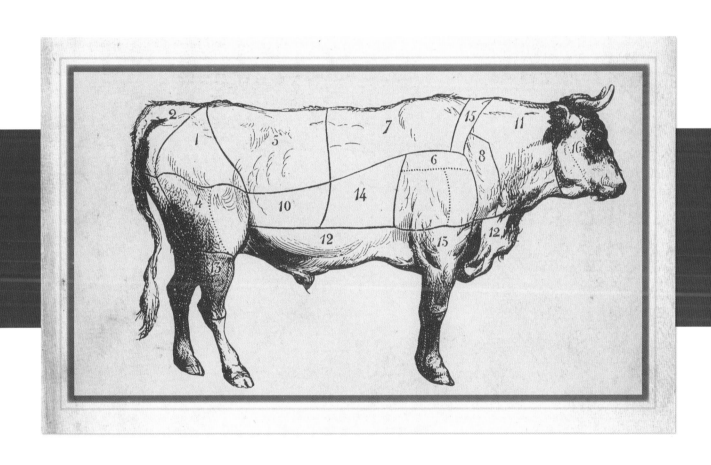

From Farm to Table

For cattle farmers, ranching is not just a business but a way of life—they love they the outdoors and they love their animals. Livestock raised in a healthy and humane environment produce the best beef for the table.

Raising Cattle

Cattle really are what they eat. Some cattle are raised on grass, other cattle eat a diet based mainly on nutritional supplements, grain (to add weight), or a mix of grass, hay, and silage. Not surprisingly, animals raised on a natural diet, in a free-range environment, taste better. Even though pasture is cheaper for the farmer, the cattle take longer to raise to maturity. But the benefits are clear on the dinner plate, although you'll also feel them in your wallet! There are two systems of farming the beef we eat—one used for dairy-beef crossbreeds and the other for beef-only herds.

DAIRY-BEEF CROSSBREEDS

Dairy-beef crossbreeds, whose mothers are dairy cows and fathers are prime beef cattle, account for about half of all beef slaughtered for meat. These intensively raised cattle spend their early lives reared in sheds and are fed milk from automatic feeders. They live the rest of their days experiencing a mix of outdoor grazing during spring and summer and intensive indoor finishing, where the cattle receive a nutritionally controlled grain diet to fatten them quickly for market. Dairy-beef cattle are naturally leaner and carry less meat than pure beef cattle. For this reason, they need to be more intensively farmed and given additive-rich feeds to bring them up to a marketable weight.

PURE BEEF CATTLE

In contrast to dairy-beef herds, pure beef cattle enjoy a more natural way of life. The calves from "suckler" herds feed from their mothers in a grazing environment for up to ten months. Depending on the quality of the pasture, the cattle go on to spend their remaining lives grazing naturally and being fed supplementary grains. Beef cattle receive a mix of grass and grain foods at different stages of their lives. They are given finishing nuts—high-grade nutritional supplements—for two–five months before they are slaughtered to increase the fat layer and marbling throughout the body.

PEDIGREE CATTLE

There's a lot of talk about superior breeds of beef cattle, but the value of pedigree has been overdone with regards to the meat. It is cynical to say that publicizing the pedigree of beef is just a sales tool, because different breeds do have slightly different qualities. However, the depth of flavor we prize so highly comes mainly from how the cattle are raised, fed, and cared for. Good butchers have a saying that "great beef is grown and not made." Provenance is vital and a reliable

Pampered Pedigree

Perhaps the most famous breed is the Japanese Wagyu (also known as Kobe beef). These cattle are highly revered, and are shrouded in myth and fantasy regarding their lifestyle—drinking beer to relax their flesh, massage, even classical music—all in the name of good beef. The result is supremely tender and heavily marbled meat that costs a fortune. Buyer beware: Genuine Kobe beef is only available in Japan.

Originally used as a working animal, white Italian Chianina cattle are prized for the quality of their meat.

a knowledgeable butcher, buy direct from the producer at farmers' markets or from a farm shop. The most famous breeds across North America and in the U.K. are Aberdeen Angus, Hereford, and Charolais. Galloway and Beef Shorthorn cattle are farmed widely in the U.S., while Lincoln Red, Red Devon, Sussex, and Dexter are herds reared only in the U.K. Around the world there are other breeds, such as Japanese Wagyu and Italian Chianina, that have gained fame for their flavor and quality.

Hanging and Aging

All beef, with the exception of veal, benefits from hanging (also known as aging) to mature the meat after slaughter. In a grown animal there is a lot of connective tissue, and hanging allows the natural enzymes in the carcass to break this down, making the tissue softer and more elastic.

DRY-AGED VERSUS WET-AGED

Hanging takes place in a walk-in cold room at the butcher's or abattoir. The carcasses are hung from ceiling racks, spaced evenly apart and surrounded by circulating air set at 34°F (1°C). This is known as "dry hanging" or "dry aging." As the meat loses moisture, it also loses weight—in a carcass hung for 28 days, 10 percent of the original weight is lost. Professionally managed dry aging is expensive and, as a result, mass-market beef suppliers resort to "wet" aging to save time and money.

Wet-aged meat is vacuum-packed straight after slaughter to prevent moisture and weight loss, therefore maximizing profits for the producer. For the discerning cook, dry-aged meat is far superior—wet-aged meat is drier after cooking because the excess water expands with heat, stretching the delicate fibers, which contract quickly on cooling, producing a tougher result. It is hard to get the all-important caramelization and seared surface with wet-aged meat, especially on steaks and roasting cuts. Dry-aged beef simply tastes much better and is worth the extra investment.

butcher will know exactly from where his or her meat comes, how it was raised, slaughtered, and handled before sale. A butcher who has real passion for the craft will answer your questions about the meat's origins without hesitation. If you do wish to buy beef from a named pedigree, then seek out

Flash Cooking

Flash cooking is just as it sounds, super speedy; there are some beef cuts that benefit from being shown the inside of a supremely hot pan for just a few seconds.

The term "minute steak" is used for fast, panfried beef simply because of the need to cook it with haste. For the ultimate result, always fry in a preheated, heavy pan or skillet for even heat distribution, adding just a little oil for flavor. Intense, high heat is the key to success—the browned color that comes from the quick caramelization of the meat makes flash-cooked cuts look and taste awesome. Any more contact with heat toughens up the meat fibers considerably, making it unpleasantly chewy and losing essential tenderness.

Fat Level

Meat fat is celebrated for the valuable moisture and texture it adds. Butchers talk of "visual leanness" as a reference to amount of fat shown. Hamburgers usually contain 20% fat, although some are prepared with 25% fat for maximum flavor.

Ground Beef

The best choice for burgers, ground beef can be made from any cut but is usually prepared with offcuts and trimmings—a way of getting the most from the animal. "Grinding" breaks down the connective tissue mechanically and tenderizes. Hamburgers take longer cooking than other flash-cooked cuts, depending on thickness. Burgers should be cooked to medium-rare and must be very fresh.

A.K.A. Mince

Hangar Steak

A long muscle that attaches to the rib cage, the hangar steak is fantastic fried, then sliced across the grain to maximize tenderness. Historically, this is a cheap cut, known as the "poor man's filet." But the hangar steak is now on-trend and the laws of supply and demand mean that it's more pricey.

A.K.A. Onglet

Skirt

A flat sheet of meat from the inside of the flank, the skirt has a coarse, open, almost-ribboned texture. Although it can be cooked slowly to tenderize, it is also great when panfried and served rare, but must be sliced thinly across the grain before serving.

A.K.A. Goose skirt, fajita meat, Philadelphia steak

Flatiron

A thin ribbon of muscle sandwiched between layers of fat and gristle close to the blade, the flatiron steak has plenty of fat marbling streaked through it to add both flavor and juiciness when cooked.

A.K.A. Top-blade steak, butler's steak

Grill & Broil

The perfectly cooked steak is the "holy grail" for the committed beef fan, and knowing your cuts is the secret to cooking them right. If in doubt, ask your friendly butcher for some professional advice.

The best steaks are cut from the center of the animal, around the ribs and down the rump. The area inside the ribs—the rib eye, sirloin, and tenderloin—doesn't do much work so the meat is soft and tender. The harder-working rump is a touch tougher, but that exertion equals extra flavor—many butchers say that a properly cooked rump steak is by far their favorite: The steak's sizzle and taste comes from the marbled fat and ensures it comes out real juicy. Some steaks have extra fat: With sirloin, it is a layer on one side; with rib eye, it's a central disc of fat. This fat is kept on the steak during cooking.

Sirloin

Sirloin is affordable, tasty, and lean. As it's tougher than other cuts, ask the butcher for a "single muscle" sirloin steak. The whole sirloin is made up of four separate muscles, held by gristle and connective tissue, so you get a better result from a single muscle.

A.K.A. Rump, popeseye (thin slice of sirloin steak)

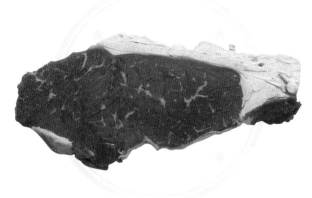

NY Strip

Cut from the lower portion of the ribs, this is a top-quality beef steak, offering a layer of creamy white fat on one side. When your skillet is searing hot, use tongs to gently hold the steak fat-side down for a minute to melt the fat into the meat.

A.K.A. Sirloin

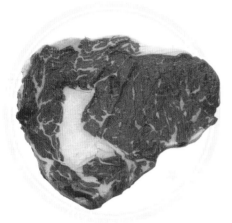

Rib Eye

The rib eye is a large, juicy, boneless prime cut packed full of flavor. The disc of white fat at the center melts slightly during cooking, and ensures that the meat is tender and succulent.

Is Rare Always Best?

Rare is not always the right choice when it comes to cooking steak. Rib eye benefits from being served medium-rare to give its central disc of tasty fat a chance to render down on cooking. But resting is the most crucial part of cooking steak as this allows the muscle fibers time to relax and the cooking juices to be absorbed back into the meat. Five minutes on a warm, foil-covered plate is the minimum resting time. If you dare to skip this step, you will be quite disappointed with the end result.

T-Bone

Weighing in at a whopping 16 ounces (450 g), the T-bone is the ultimate "man-sized" steak. Cut from the short loin, it is made up of a sirloin steak on one side of the bone and a piece of fillet on the other. While it looks impressive on the plate, this steak can be a tricky one to cook perfectly as the two different cuts benefit from different cooking times.

Tenderloin

Carefully cut from the spine, this expensive steak is less than one percent of the animal, so no wonder it's so pricey! Tenderloin is split into separate cuts: Filet mignons are from the thin end, while larger Chateaubriand steaks come from the wider end.

A.K.A. Filet, eye filet

Roast

Here's where you let the oven take the strain and make the meal for you. Making roast beef is genuinely simple, but always choose the right cut to ensure the result is tender and succulent.

Well-roasted beef is a beautiful sight to behold. This is meat-eating at its purest and most unadulterated and, for that reason, the quality of the beef is vital. With a roast, there is little place to hide inferior meat behind a mask of other flavors, so it pays to buy the best you can afford.

For maximum flavor, always start your roast with a little caramelization on the outside, either by searing in a very hot pan or by roasting in a sizzling, preheated oven for the first 30 minutes. At the end of cooking, for truly tender, juicy meat, rest the roast—covered—for a minimum of 15 minutes.

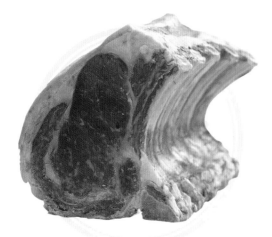

Standing Rib Roast

Perhaps the ultimate roast, this cut has it all— perfectly marbled meat and a natural layer of creamy fat to self-baste, plus it's cooked on the bone for maximum flavor: a little like a beef rack of lamb. This cut, boned and sliced, makes rib eye steaks.

A.K.A. Rib roast, prime rib, forerib

Sirloin Roast

Leaner than a rib, this marbled cut is very tender and comes with a layer of fat to baste as it cooks. A skilled butcher can cut the flank of the sirloin wider, then trim so that the sirloin is wrapped in its own fat. Pre-packed roasts may use reconstituted fat.

A.K.A. Rolled loin roast

Round Roast

Round cuts include a variety of fantastic roasting options, including bottom round (left). Because these cuts are lean, it is usual to roll and tie the meat with a thin layer of extra fat to keep it from drying out. The specific names of round roast cuts vary according to how the animal is butchered. This part of the animal may also be used to make corned beef.

A.K.A. Top rump, round tip, topside, silverside, eye round

Other Roast Cuts

- **ROLLED RIB** This is the same as a rib roast, but has the bones removed, then rolled and tied for convenient slicing after roasting. Rolled rib eye is the most tender rib roast because it is from the very center of the animal, but has a little less flavor than a traditional bone-in roast.

- **TENDERLOIN** Usually reserved for individual steaks, the entire central tenderloin can be roasted as a stunning centerpiece. The cut constitutes the long muscle that runs along the spine of the animal. It is the most expensive—and most tender—of all roasting cuts.

- **SHELL ROAST** This expensive sirloin roast is taken from the short loin, without any of the tenderloin. It is sold boneless or bone-in, and is sometimes called a striploin roast.

Braise & Stew

Braising and stewing are essentially the same thing: the gentle cooking of meat—long and slow—in some sort of liquid. Wine, stock, vegetables, herbs, and selected spices add extra flavor dimensions.

While the best meat for frying and grilling comes from the tender middle, the cuts that are prime for slow braising and stewing come from the hardworking front and back—the forequarters and hindquarters of the animal. The lean, sinewy muscles of the legs, shoulder, and flank can be stringy and tough if not properly cooked, but produce the most unctuous and melting-fleshed dishes when treated with patience and respect. Slow cooking is an essential tool in the thrifty cook's repertoire, these cuts being by far the most economical to buy.

Shank

The meat from the legs is relatively low in fat as these muscles are the working powerhouse of the animal. This cut contains a mass of connective tissue that rewards the patient cook plentifully with rich juices and meltingly fork-tender meat.

A.K.A. Shin, leg

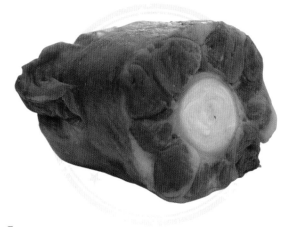

Oxtail

Single pieces (above) or thick rings of sliced oxtail, studded with bone, marrow, and gelatinous tissue produce the richest and most luxurious stew you will ever taste—one certainly belying the supreme economy of the meat.

Brisket

This homely cut comes from the breast section—the underside of the forequarter—and is layered with smooth fat (known as the "fat cap"). The fat adds flavor and tenderness to the slow-cooked meat. This cut is also used to make corned beef.

Short Rib

Larger and meatier than their pork counterpart (sparerib), this cut has the added bonus of being cooked on the bone, really intensifying the flavor of liquid sauces during cooking. The rib bones themselves slip out with ease after slow cooking.

A.K.A. Thin rib

Stew Meat

Coming from a really hardworking part of the animal, there is plenty of connective tissue that breaks down over time to produce an rich, unctuous sauce. A good option for marinating.

A.K.A. Braising steak, shoulder

Stewing and Pot-Roasting

A piece of advice for slow cooking—buy your beef in one piece so you know it comes from the same part of the same animal, before preparing it yourself or asking your butcher to cut it. Packaged "stewing" meat may come from the forequarters or hindquarters, which have different cooking needs.

Less common cuts are worth hunting out for their intense flavor. Bavette or flank are super-satisfying when given the slow treatment, as is skirt—the long muscle from around the diaphragm. To maximize tenderness, slice across the grain before cooking. Top round (silverside) is another classic pot-roast cut.

Veal

Ethically reared "rose" veal has a supremely delicate flavor and an ever-so-soft texture. As an alternative to standard beef, veal offers its own range of special recipes, especially those from Italy.

V eal is the meat from calves or young cattle, and it is essentially a byproduct of the dairy industry. In order for a dairy cow to stay in milk production she needs to give birth to a calf every year. However, the majority of male calves are surplus to requirements and are shot or sent to slaughter. Due to poor animal welfare standards in the past, veal has had a controversial history. Due to strong consumer demand, ethically reared veal—rose veal—is now available, affordable, and enjoyed by all. The reality is that veal are slaughtered at an older age than chickens, lambs, or pigs.

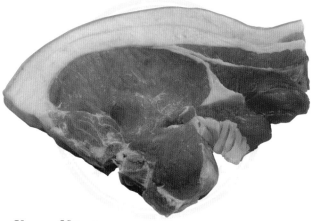

Loin Chop

This is the veal equivalent of a T-bone or Porterhouse steak. It is best broiled or panfried medium-rare over a high heat just as you would a steak. Veal is naturally low in fat and has virtually no marbling, so butter or oil are essential for adding moistness and flavor.

Shank

This is the cut used for the Italian classic dish, "osso buco." The shank is cut into thick slices across the bone so that each piece contains a cross-section of pink meat and white bone, complete with the sweet internal marrow.

A.K.A. Fore shank, hind shank, shin, knuckle

Scallop

Normally cut from the round end (rump), these thin slices are often breaded and fried or prepared in the Italian style, "saltimbocca." Left whole, the entire round can also be roasted. Ask your butcher what cuts he can acquire for your chosen recipes.

A.K.A. Escalope

Liver

Said to be the most tender of all livers, shiny, smooth-textured calf's liver is classically sliced and panfried with onions and pancetta. It is remarkably nutritious and rich in essential vitamins and minerals. This essential organ cut must be purchased very, *very* fresh from a reliable source.

Raising Veal

Old-school, mass-produced "white veal" comes from purely milk-fed calves who have spent their short lives tethered in a crate unable to move for fear of too much muscle development. White veal also indicates a lack of iron in the meat and a restricted diet for the animal. Crate-rearing is now banned in many countries.

Unacceptable rearing methods have been replaced by a new wave of ethically reared rose veal, so called because of its pink color that comes from freedom of movement, access to sunlight, and a varied diet of milk, grass, and cereals. In fact, consuming rose veal is regarded as a positive environmental choice for meat-eaters because it provides an economic and culinary use for the unwanted male offspring of dairy cows, whose lives are usually ended a few days after birth.

Rose veal is slaughtered at six–eight months of age, meaning that they can—arguably—be classified as young beef cattle rather than as baby calves.

Perfect Porterhouse steak

The Porterhouse is just a very large T-bone—the first three or four steaks off the end of the short loin. This means that you can cook a T-bone using the same method, just more briefly.

Serves 2

Ingredients

2 Porterhouse or T-bone steaks, about 1 inch
  (2.5 cm) thick, at room temperature
salt and black pepper
2 tablespoons (30 ml) vegetable or olive oil
¼ stick (30 g) butter

Butcher's Tips

If you salt the steak just before cooking, there is no time for the essential meat juices to be drawn out and evaporate in the pan. The salt rests on the steak, leaving the juices inside, resulting in a clean, hard, seared surface. However, if you salt the meat for a much longer period of time (40 minutes or even more) most of the juices are drawn out, but then have time to be absorbed back into the meat. Cooking the steak for between three and 40 minutes after salting is the worst way to cook the meat because liquid builds up on the outside of the meat and also in the pan. This makes it difficult to get a good crust.

Method

1 Season the steaks with salt and pepper just before cooking. Heat the oil in a large, heavy frying pan or griddle until it smokes.

2 Hold each steak upright in the pan with the fatty side touching the hot surface. Allow some of the fat to melt, then lay the steak flat. Cook over medium heat, flipping occasionally. T-bones can be tricky to cut straight, so be aware that uneven thicknesses will affect the cooking times. Cook for 6–8 minutes—medium-rare is best for this cut— depending on thickness. To test, insert a meat thermometer into the steak, away from any bone, fat, or gristle: It should read 135°F (58°C). You can also use the "prodding" method, used opposite.

3 Add the butter to the pan. Allow to melt and sizzle, then remove the steak from the pan.

4 Cover the steak and allow the meat to rest for 5–10 minutes before serving. The meat will finish cooking during this time.

Perfect sirloin steak

Bring the steakhouse atmosphere home with a good red slab of sirloin. Serve with a heap of potatoes, done any way you like, along with your favorite salads and sauces.

Serves 2

Ingredients

2 sirloin or rump steaks, about ¾ inch (2 cm) thick, at room temperature
2 tablespoons (30 ml) vegetable or olive oil
salt and black pepper

Chef's Tips

To check the pan is hot enough for your steaks, hold your hand over the top. If it feels too hot to hold your hand in place, remove the pan from the heat for a few seconds. If the pan heat just feels warm, increase the heat. If the pan is not hot enough at the start, the steak will stew and turn tough. If it's too hot, the meat will burn. The meat should sizzle as soon as it goes in the pan.

Method

1 Brush the steaks with oil and season with salt and pepper. Heat a large, heavy frying pan until it starts to smoke. Add the steaks and cook over medium heat on both sides.

2 Use the "prodding" method to cook the steak for the right time, and to test for doneness. Here's how: press your thumb and forefinger together, then feel the firmness of your flesh at the base of the same thumb. Move your thumb from your first finger to your pinky, and you'll notice that your thumb flesh feels harder as the muscle tenses, just like a steak as it cooks. For rare steak, cook for 2½ minutes each side, then press the meat. It should feel soft, just like your thumb flesh when the forefinger and thumb pad are held together. For medium–rare, cook for 3 minutes on each side. The meat will feel harder, like your thumb flesh when the middle finger is pressed to the thumb. For well-done, cook for 5 minutes each side, or until the meat feels firm, like your thumb flesh when the thumb pad presses the little finger.

3 Let the steaks rest, covered, for 5–10 minutes before serving. They will become more tender as they rest.

Hangar steak with peppercorn sauce

The hangar steak "hangs" from the last rib, near the kidneys, giving it a deeper flavor. Cream gives the sauce added richness.

Serves 4

Ingredients

1 whole hangar steak, trimmed
¼ cup (60 ml) vegetable oil
1 onion, peeled and finely chopped
1 clove garlic, peeled and finely chopped
1 tablespoon (10 g) pink peppercorns
1 tablespoon (10 g) black peppercorns, coarsely cracked
½ cup (120 ml) red wine
¼ cup (60 ml) heavy cream
½ cup (120 ml) chicken or beef stock
salt

Butcher's Tips

To prepare an untrimmed steak, first cut off the membrane and fat from the outside, leaving a central "feather" of sinew. With your knife pressed against the sinew, scrape the meat off the sinew to give one "clean" piece of meat, and another piece with sinew. Lay the steak, sinew-side down, and scrape away the remaining sinew as if skinning a fish.

Method

1 Cut the steak into 4 portions and tap each one gently with a meat mallet. Set aside.

2 Heat half the oil in a pan. Add the onion and garlic, and fry gently until softened. Add the pink peppercorns, black pepper, and wine. Simmer until most of the liquid has evaporated. Stir in the cream, stock, and salt to taste. Simmer until thickened, stirring occasionally.

3 For the steak, heat a large, heavy frying pan over medium heat. Brush the steaks with the remaining oil and season with salt. When the pan is smoking, add the steak and fry for 1½ minutes on each side for medium-rare. Fry the steak in batches if they cover more than half of the pan's surface. Remove the steak and allow to rest, covered, for 3 minutes. Cook the remaining steaks.

4 Add the sauce to the meat pan and stir to deglaze the juices. Stir in any juices from the rested steak. Serve the steak with the sauce on top.

Sirloin with sauce verte

A favorite restaurant cut, prime top sirloin offers a tight texture
and a clear marbled grain—tender, with a robust flavor. A fresh
herb sauce offers a refreshing contrast to the meat.

Serves 4

Ingredients

4 prime top sirloin steaks, about 10 ounces
   (275 g) each
salt and black pepper
vegetable or olive oil, to grease the grill

For the sauce
½ cup (75 g) flat-leaf parsley, loosely packed
½ cup (75 g) basil, loosely packed
½ cup (75 g) mint, loosely packed
1 tablespoon (15 ml) Dijon mustard
3 anchovy fillets
2 cloves garlic, peeled and minced
2 tablespoons (20 g) capers
1 teaspoon (5 g) ground black pepper
good pinch of cayenne pepper
2 tablespoons (30 ml) lemon juice
salt
½–¾ cup (125–175 ml) extra-virgin olive oil

Method

1 Season the steaks on both sides with salt and
   pepper. Grill or broil on an oiled grill rack until
   rare or medium-rare, according to taste. Cover
   and allow steaks to rest.

2 Meanwhile, put all the sauce ingredients, except
   the oil, into a food processor. Process until
   coarsely chopped.

3 With the food processor running, gradually
   drizzle the oil in a slow, steady stream until
   fully incorporated.

4 Serve the steaks with the sauce spooned on top.

Bistecca alla fiorentina

Known affectionately as a Tuscan Porterhouse, the fame of this Florentine steak rests on the quality of the beef as the only flavorings are bay, rosemary, and peppercorns.

Serves 4

Ingredients

2 pounds (900 g) T-bone or Porterhouse steak
  on the bone
arugula (rocket) leaves, to garnish
Parmesan flakes, to garnish

For the marinade
coarsely ground black peppercorns
coarsely ground pink peppercorns
1–2 bay leaves
1 sprig rosemary
olive oil, to drizzle

Butcher's Tips

Depending on the size of the butcher's cuts, you may need to use two or three separate steaks for four servings. Rib eye steaks can also be used, but cook over medium heat to ensure that the disc of fat at the center of the steak starts to melt. A fresh lemon wedge makes an excellent partner to the steak.

Method

1 Place the meat in a large, non-metal dish. Season both sides of the steak with the marinade ingredients, and rub in with your fingers. Cover and refrigerate for at least 5 hours, or overnight.

2 Allow the meat to return to room temperature before you begin to cook. Brush a large, heavy frying pan with a little olive oil, then heat until smoking. Meanwhile, warm an ovenproof plate or dish in the oven.

3 Cook the steak for 2 minutes on each side, or until just rare.

4 Transfer the steak from the pan to the heated plate or dish. Cover with foil and allow the meat to rest for 5–10 minutes before serving. The meat will continue to cook and tenderize while it rests.

5 Serve the steak garnished with arugula (rocket) leaves and flakes of Parmesan cheese.

Veal medallions with morels

For this exquisite starter, tender morsels of veal are married with a creamy sauce of wild mushrooms and offset by deep-green ramps and watercress.

Serves 4

Ingredients

6–8 ounces (175–225 g) veal medallions
salt and black pepper
1 stick (115 g) butter
2 cups or about 8 ounces (225 g) morel
   mushrooms (or any other mushroom variety)
2 cups or about 8 ounces (225 g) ramps or
   scallions, sliced
½ cup (120 ml) heavy cream
2–3 cups or about 8 ounces (225 g) wild or
   farmed watercress

Method

1 Preheat oven to 400°F (200°C). Season the veal with salt and pepper. Melt 2 tablespoons of the butter in a pan over medium heat. Add the veal and sauté until nicely browned on both sides.

2 Transfer the veal to the oven and cook for 3–4 minutes, or until the internal temperature of the meat is 135°F (57°C) for medium-rare. Remove the veal from the oven, cover, and allow to rest.

3 For the sauce, melt 2 tablespoons butter with salt and pepper to taste. Add the mushrooms and sauté for 4–5 minutes, or until soft. Remove from the pan, cover to keep warm. Melt 2 more tablespoons butter, then add the ramps or scallions. Sauté over low heat for 4–5 minutes, until softened but not brown.

4 Stir the cream into the ramps, and bring to a gentle simmer to thicken slightly. Stir in the watercress and the last 2 tablespoons butter. Check the seasoning again with salt and pepper. To serve, spoon the cream sauce onto warmed plates. Top with the veal medallions and the sautéed mushrooms.

Saltimbocca alla romana

Only the charm and confidence of Roman cuisine can offer the world a dish that means "jumps into the mouth." Fresh sage and ham are all that is needed to season this delicate veal cut.

Serves 4

Ingredients

4 veal scallops (topside or escalope), about 4 ounces (115 g) each
4 large sage leaves or 8 smaller leaves
4 slices Parma ham or prosciutto
1 tablespoon (15 g) all-purpose flour, to dust
¼ stick (30 g) butter
4 tablespoons (60 ml) white wine or marsala
salt and black pepper

Method

1 Use the palm and heal of your hand to flatten the slices of meat until they are very thin. Veal is delicate so be careful not to press too hard or it may tear apart. Lay the sage leaves in the center of each slice.

2 Place the slices of ham over the sage and use your hand to press the layers together.

3 Put a little flour onto a plate. Carefully dust both sides of the layered veal in the flour.

4 Melt the butter in a large, heavy frying pan over medium heat. Add the veal, with the ham facing down. Cook for 20 seconds, or until lightly browned. Carefully turn over the veal, and fry again on the other side for just 20–30 seconds, or until just cooked.

5 Drizzle the wine into the pan and stir over medium heat until the sauce bubbles. Season with pepper. Only add salt at the table because Parma ham is already very salty. Serve with the deglazed juices from the pan.

Steak salad with arugula

This show-stopper of a salad puts a great steak at center stage.
Mop up the intensely flavored juices with some crusty bread
and enjoy with a perky little Pinot Noir.

Serves 4

Ingredients

¼ cup (60 ml) olive oil
2 teaspoons (10 ml) balsamic vinegar
½ teaspoon Dijon mustard
salt and black pepper
4 cups (250 g) arugula (rocket), rinsed and dried
2 tablespoons (25 g) pine nuts
¼ stick (30 g) butter
2 sirloin (rump) steaks, about 10 ounces (275 g) each
1 scallion (spring onion), trimmed and sliced

Method

1 In a small bowl, whisk together 2 tablespoons
   olive oil with the vinegar and mustard. Season
   with salt and black pepper, then set aside. Divide
   the arugula (rocket) among four plates.

2 In a heavy skillet, lightly toast the pine nuts by
   shaking constantly over high heat for a minute
   or two. Scatter the nuts over the arugula (rocket).

3 Place the butter in the skillet and return it to the
   heat. Add the remaining oil and swirl to combine.
   When hot, lay the steaks in the skillet and sear for
   4 minutes, until browned, spooning over the fat
   occasionally. Turn the steaks and cook for another
   3–4 minutes for medium-rare. Transfer the meat to
   a cutting board or warmed plate, cover, and allow
   to rest for 5–10 minutes.

4 Scatter the scallions over the arugula (rocket).
   Slice the steak thinly and place on top of each
   salad. Drizzle with the dressing and serve.

Steak with jalapeño salsa

For this seared steak, ask your butcher for a slab of skirt steak.
The cut is quite a rarity since there's only one per animal, but
it's a good, long piece of meat that is so easy to grill or fry.

Serves 4

Ingredients

1¼ pounds (500 g) skirt steak
salt and black pepper
2 tablespoons (30 ml) olive oil

For the sauce

1 cup (100 g) cilantro (coriander) leaves
2 cloves garlic, peeled
1 jalapeño pepper, seeded
2 scallions (spring onions), chopped
1 tablespoon (10 g) coarse sea salt
juice and zest of 1 lime
1 cup (240 ml) olive oil
salt and black pepper

Method

1 Season the steak on both sides with salt and
pepper. Set aside.

2 For the sauce, place the cilantro (coriander),
garlic, jalapeño, scallions (spring onions), salt,
lime juice, and lime zest in a food processor.
Pulse the ingredients until evenly chopped, but
still granular in texture. Scrape the mixture into
a bowl and stir in 1 cup (240 ml) olive oil. Season
with salt and pepper.

3 In a large sauté pan, heat 2 tablespoons olive
oil over medium heat until smoking hot. Add
the steak and sear for 2–4 minutes on each side,
depending on thickness. For a medium-rare steak,
the meat should give slightly when prodded
gently. Remove the steak from the pan, cover, and
allow to rest for 5–10 minutes.

4 Using a very sharp knife, slice the steak against
the grain and serve with the sauce.

Dave's best beef burger

Pick up some fatty ground beef from your butcher, and add select seasonings to make, possibly, the world's best burger. The fat keeps the burgers juicy—this is no place for lean cuisine.

Makes 4 burgers

Ingredients

1 small red onion, peeled and grated
1¼ pounds (500 g) ground beef, with at least 20% fat
1 teaspoon (5 ml) honey
1 teaspoon (5 ml) Worcestershire sauce, optional
2 tablespoons (20 g) chopped fresh parsley
½ teaspoon paprika
generous pinch red pepper flakes
black pepper
2 tablespoons (30 ml) olive oil

To serve
burger buns, red onion slices, sliced tomatoes, and
  shredded lettuce

Butcher's Tips

Salting the burger mix can draw out essential juices and make the mixture fall apart. Season with salt after cooking, if necessary.

Method

1 For the burgers, put the grated onion into a strainer over a bowl to let the juices drain away. Mix the onion with the other burger ingredients, except the oil, until well combined. Using wet hands, form the mixture into patties, about ¾ inch (2 cm) thick. Chill for at least 15 minutes to prevent the patties from falling apart.

2 Heat the oil in a large frying pan. Add the patties, and fry over high heat for 3–4 minutes on each side, or until tender and cooked through. Do not be tempted to press the patties with your spatula.

3 Split the buns and toast them lightly. Fill with the hot burgers, red onion, tomatoes, lettuce, plus your favorite condiments.

Salt beef & red onion marmalade sandwich

Salting turns a really tough cut into something super tender—just clear some room in your fridge so the brisket can sit and soften for a couple of weeks.

Serves 4

Ingredients

For the salt beef

2 cups (475 ml) water
3 cloves garlic, plus 1 whole head garlic
¾ cup (110 g) sea salt
1 teaspoon (5 g) peppercorns
6 sprigs thyme
1¼ pounds (500 g) boneless beef brisket, trimmed
1 carrot, peeled and chopped
1 leek, peeled and chopped
1 onion, peeled and chopped
toasted bread and arugula (rocket), to serve

For the marmalade

1½ pounds (680 g) red onions, peeled and sliced
¼ stick (30 g) butter
½ cup packed (120 g) dark brown sugar
1 stick cinnamon
1 star anise
juice and zest of 1 orange
salt and black pepper
⅓ cup (80 ml) balsamic vinegar

Method

1 For the salt beef, pour the water into a pan. Bash three garlic cloves to split open, and add to the water with the salt, peppercorns, and thyme. Bring to a boil. Cool liquid, then pour over the beef, adding extra water to cover. Set a plate and a weight over the beef to keep it submerged. Refrigerate for 2 weeks, turning the meat every other day.

2 After 2 weeks, remove the beef and wash in fresh water. Put into a stainless steel pan and cover with cold water. Cut the garlic head in half horizontally, and add to the pan with the carrot, leek, and onion. Simmer for 2–3 hours, skimming the scum from the surface as it rises. The beef is cooked when it is soft and falling apart. Allow to cool, then remove the beef and refrigerate.

3 For the marmalade, sweat the onions in the butter for 15 minutes, or until soft. Add the remaining ingredients. Cover the mixture with a circle of parchment paper, then simmer for 2 hours, or until thick. Cool, then spread on toast and serve with sliced beef. Garnish with arugula (rocket).

Nose to tail eating: organ meats

Proponents of the Paleo diet have put offal back on the menu, so expand your horizons beyond calf's liver: Really fresh offal provides great eating.

HEART

Heart is a hard-working muscle that is very lean, except for an outer layer of fat. You can either slice and flash-fry; quickly broil, as you would a steak, or stuff it and cook it slowly to break down the muscle. In the classic Peruvian dish, anticuchos, heart is sliced thinly and cut into squares, then marinated before being threaded onto skewers and grilled.

KIDNEYS

In Britain, steak and kidney pie and deviled kidneys (once a common breakfast dish) are popular. Or, head farther afield to the Szechuan region of China where spice-dusted kidneys (usually lamb) are served hot off the grill. The trick is to cook these delicate morsels quickly and serve them pink. They need to be ultra-fresh or they will taste bitter, so cook on the day you buy them.

SWEETBREADS

Delicate and creamy when cooked, sweetbreads are the thymus and pancreas glands of calves and lambs. Soak in several changes of cold water for a few hours before cooking. Sweetbreads are usually poached in stock then breaded and fried. After poaching, cut off any gristle or membrane.

TRIPE

Made from the first three chambers of a cow's stomach, tripe is rubbed with rock salt, rinsed, soaked in hydrogen peroxide, then rinsed several times until it is snowy white. It should

Fries With Everything

Rocky Mountain oysters or fries in the U.S., prairie oysters in Canada, *huevos de toros*—literally "bull's eggs" in Spanish—testicles are most often known by a coy nickname. Although they're called bull's testicles, ask your butcher for calf's testicles, tender nuggets about the size of a walnut. They need to be skinned before cooking and are usually served breaded and fried with a spicy dipping sauce.

be clean and odorless. Once prepared, tripe is stewed in an aromatic broth until tender. In Florence, Italy, it is so popular, that stewed tripe is served from street carts in a bun with garlic and parsley sauce.

TONGUE

Preparing tongue can be a chore, but this delicious and low-priced meat is worth the trouble. First scrub the meat with a vegetable brush, then soak in several changes of cold water. Next, simmer gently with herbs and spices until soft. Finally, pull off the skin, and remove any fat and gristle. Now you can add the sauces and flavorings that make people love this rich meat. Try tongue tacos, or eat it sliced cold in sandwiches.

▶ Tripe in all its forms provide the makings of a hearty dish and is popular in all traditional meat-eating cuisines.

Brioche-crusted sweetbreads with leeks & truffle sauce

These mini mouthfuls of velvety veal organ meats are encased in golden crumbs and served with delicate slivers of leek and a classically French Madeira–mushroom sauce.

Serves 4

Ingredients

1 pound (450 g) veal sweetbreads
½ cup (60 g) all-purpose flour, seasoned with
   salt and black pepper
3 eggs, beaten
2 cups (120 g) fresh brioche crumbs
¼ stick (30 g) butter

For the leeks
½ stick (60 g) butter
1 large leek, washed and cut into short, thin strips
¼ cup (60 ml) water
salt and black pepper

For the truffle sauce
¼ stick (30 g) butter
1 shallot or white onion, peeled and finely sliced
2 cloves garlic, peeled and smashed
1 sprig thyme
½ cup (about 50 g) mushrooms, sliced
1 cup (240 ml) Madeira wine
2 cups (475 ml) brown chicken stock
3 tablespoons (45 g) truffle butter

Method

1 Soak the sweetbreads in water for 24 hours in the refrigerator. Drain, then blanch the sweetbreads in boiling water. Transfer to cold water, then drain again. Pull off the veins, gristle, and membrane.

2 Preheat oven to 350°F (180°C). Cut sweetbreads into 4 equal slices. Dust in seasoned flour. Dip in beaten egg, then dredge in brioche crumbs to coat. Melt the butter in a frying pan, and sauté the sweetbreads until golden. Bake for 20 minutes, or until the internal temperature is 140°F (60°C). Set aside, but cover to keep the sweetbreads warm.

Chef's Tips

Only use genuine truffle butter—made with earthy white or dark truffles—rather than brands containing synthetic ingredients. If you can't find the genuine article, use regular butter and add a few rehydrated wild porcini mushrooms for extra flavor.

3 While the sweetbreads are baking, prepare the leeks. Melt the butter in a pan. Add the leeks, water, and seasoning. Cook gently until the leeks are soft and the liquid has evaporated. Remove from the heat and cover to keep warm.

4 For the truffle sauce, melt the butter in a pan. Add the shallot or onion, garlic, thyme, and mushrooms. Sauté for 3–4 minutes, or until softened. Add the wine and simmer until most of the liquid has evaporated. Add the stock and simmer until the liquid is thick enough to coat the back of a spoon. Strain the sauce through a fine strainer, then return to the pan. Add the truffle butter and swirl until melted. Season to taste.

5 To serve, place the leeks in the center of warmed plates. Top with the crispy sweetbreads and surround with the truffle sauce.

Sweetbread fritters on banana & apple purée

Japanese panko bread crumbs make these fried sweetbreads extra crispy. Served with a sweet-tart fruit purée, they make an impressive first course.

Serves 4

Ingredients

For the fritters
1 pound (450 g) veal sweetbreads
2 cups (475 ml) chicken or vegetable stock
all-purpose flour, to dust
salt and black pepper
2 eggs, beaten
panko bread crumbs, to coat
vegetable oil, to fry

For the purée
3 large Bramley's Seedling apples, or other
   soft-fleshed cooking apple
juice of 1 lemon
2 tablespoons (30 ml) water
2 ripe bananas

Method

1 For the fritters, soak the sweetbreads in water for 24 hours in the refrigerator. Drain, then simmer the sweetbreads in the stock for 5 minutes. Drain, cool, then carefully pull off the veins, gristle, and exterior membrane.

2 For the purée, peel, core, and chop the apples. Put into a pan with the lemon juice and water. Cover and cook gently for 10 minutes, or until soft. Mash the bananas. Add to the apples and blend until very smooth.

3 Dry the sweetbreads on paper towel. Cut into 8 pieces. Season the flour with salt and pepper. Dip the sweetbreads into the flour, then into the beaten egg. Roll them in the bread crumbs. Fry in hot oil until golden brown on all sides. Serve with the apple purée.

Seared calf's liver with anchovies

A sharp, anchovy-laced gravy offsets the velvety richness of calf's liver. Serve with ultra-smooth mashed potato and fresh green vegetables for a dinner to die for.

Serves 2

Ingredients

2 tablespoons (30 ml) vegetable oil
6 thin slices calf's liver, trimmed of sinew
salt and black pepper

For the gravy
1 tablespoon (15 ml) tomato paste (purée)
1 tablespoon (15 ml) whole grain mustard
1 tablespoon (15 ml) Worcestershire sauce
2 tablespoons (30 ml) dry sherry
6 anchovy fillets
2 cups (475 ml) fresh veal or chicken stock

Butcher's Tips

Ask your butcher for a bag of veal bones to make stock. Simmer in water with herbs, peppercorns, onions, carrots, and celery for 4 hours. Skim regularly.

Method

1 Blend the gravy ingredients, except the stock, until smooth. Set aside. Simmer the stock for 10–15 minutes, or until reduced by about half.

2 Heat the oil in a heavy skillet or frying pan. Add the liver and sear for 2 minutes on each side, until evenly colored and medium-rare inside. Remove from the pan, cover, and set aside.

3 Lower the heat under the pan. Add the gravy mixture and warm through. Whisk in the stock and simmer until reduced to a thick, pouring consistency. Serve with the liver.

Cottage pie

With all the classic flavors of the rural English countryside, this famous family dish welcomes everyone to the dinner table on a cold winter's night.

Serves 4

Ingredients

¼ stick (30 g) butter
2 onions, peeled and finely chopped
1 carrot, peeled and finely chopped
1 stick celery, finely chopped
1½ pounds (680 g) ground (minced) beef, with 20% fat
salt and black pepper
1 teaspoon (5 g) fresh thyme, optional
1½ cups (350 ml) beef stock
dash of Worcestershire sauce, optional
2–3 tablespoons (30–40 g) sweet corn or
    garden peas, optional

For the topping

2¼ pounds (1 kg) potatoes, peeled and chopped
½ stick (60 g) butter
salt and black pepper
1–2 tablespoons (15–30 ml) milk or cream

Butcher's Tips

Ask your butcher to supply beef that is coarsely ground rather than finely ground. You can also use leftover roast beef, but this should be chopped by hand rather than ground so that it does not turn into a gooey paste when cooked with the vegetables.

Method

1 Melt the butter in a large pan. Add the vegetables and sweat over low heat for 10 minutes, or until softened. Stir in the meat and cook until evenly browned. Season to taste, then add the thyme, beef stock, and Worcestershire sauce, if using. Simmer for 30–40 minutes, or until the vegetables are soft and the stock has reduced. Gently stir in the corn or peas, if using.

2 While the meat is cooking, preheat the oven to 350°F (180°C). To make the pie topping, boil the potatoes in water for 10–15 minutes, or until soft. Drain. Add half of the butter and the seasoning, then mash until smooth. If the mixture is very thick, stir in a little milk or cream.

3 Transfer the meat mixture into a baking dish. Spoon or pipe the mashed potato over the top. Top with the remaining butter, chopped up into pieces.

4 Bake the pie for 30–40 minutes, or until the potato topping is crisp and golden.

Old-fashioned meatloaf

This economical weekday standby takes only a couple of minutes to mix up and pop into the oven. Sliced, the cold leftovers make delicious sandwiches.

Serves 4

Ingredients

1 onion, peeled and chopped
1 carrot, peeled and sliced
2 cloves garlic, peeled and smashed
1 teaspoon (5 g) fresh thyme
½ cup (120 ml) tomato ketchup
1 tablespoon (15 ml) tomato paste (purée)
2 tablespoons (30 ml) Worcestershire sauce
1½ cups (about 210 g) bread crumbs
2 large eggs
1½ pounds (680 g) ground beef
3 thick slices bacon

Method

1 Preheat the oven to 375°F (190°C).

2 In a food processor, pulse the onion, carrot, garlic, and thyme together until finely minced. In a mixing bowl, stir together the vegetables with the remaining ingredients, except the bacon, until just combined.

3 Transfer the mixture to a nonstick loaf pan. Lay the bacon slices over the top, overlapping them slightly.

4 Bake for 1 hour, or until the bacon is crispy and the meatloaf is cooked through. Strain off the excess fat by carefully tilting the loaf pan up and over. Turn out the meatloaf and serve sliced with mashed potatoes and gravy.

Classic beef stock

Shop-bought stock can never compare to the real thing made at home. Your butcher will gladly supply you with a batch of fresh bones, from which you can make this essential essence.

Makes about 2 quarts (2 liters)

Ingredients

2¼ pounds (1 kg) beef bones
2 carrots, peeled and coarsely chopped
2 onions, peeled and quartered
2 sticks celery, coarsely chopped
1 tablespoon (15 ml) vegetable oil
8 peppercorns
2 bay leaves
3–4 large sprigs parsley
1 sprig thyme
salt

Butcher's Tips

Choose meaty bones with as much flesh as possible, as these make the best stock. Cheap cuts of beef, or even ground meat, can be added for extra flavor.

Method

1 Preheat oven to 400°F (200°C). Put the bones in a roasting pan and bake for 45 minutes, basting occasionally, until well browned. Toss the carrot, onion and celery in the oil and roast in a separate roasting pan until browned, but not burned.

2 Transfer the bones and vegetables to your largest pan. Add the peppercorns, bay leaves, parsley, and thyme. Cover with water.

3 Bring the stock to a boil and use a spoon to skim off any scum from the surface. Cover and simmer gently for 4 hours, skimming occasionally. Season to taste with salt. Strain the stock into a large bowl and allow to cool. Chill overnight, then skim off the hardened fat on the surface.

4 Use the stock within 3 days or freeze for up to 1 month. To freeze, reduce the stock by at least half by boiling vigorously in a pan. Cool completely, then carefully pour into ice-cube trays and freeze. Use the cubes as necessary by adding straight into your soups, stews, and sauces.

Beef & carrot stew

Long, slow cooking brings out the sweetness of cubed braising beef and soft, tender carrots. Red wine and calf's foot add a depth of saucy richness that never fails to satisfy.

Serves 4

Ingredients

¼ stick (30 g) butter
1 tablespoon (15 ml) olive oil
2¼ pounds (1 kg) braising steak, cut into
   chunks, about 2×1 inches (5×2.5 cm)
3½ ounces (100 g) bacon, cut into strips (lardons)
2 onions, peeled and sliced in rings
2¼ pounds (1 kg) carrots, peeled and cut
   into 1-inch (2.5-cm) rounds
½ calf's foot, rinsed (optional)
¼ cup (120 ml) red wine
¼ cup (120 ml) water
2 tablespoons (30 ml) brandy
salt and black pepper
1 bouquet garni (or 2 bay leaves, 1 teaspoon
   dried thyme, and 1 teaspoon marjoram)
chopped fresh parsley, to garnish

Chef's Tips

Calf's foot gives the sauce a deep, rich flavor and velvety, gelatinous texture, although the stew is still very good without it. When cold, the stew gravy sets to form an intense savory jelly that is delicious served warm or cold with soft rye bread.

Method

1 Heat the butter and oil in a large, heavy pan over high heat. When the pan is hot, add the steak and brown evenly in batches. The batch of browning meat should not cover more than half the pan. Do not crowd the pan, or the steak will steam. As each batch browns, remove and set aside.

2 Add the bacon, and cook, stirring, for 2 minutes. Lower the heat, add the sliced onions, and cook, stirring occasionally, for 10 minutes, or until the onions soften.

3 Return the steak and any juices to the pan, along with the carrots and calf's foot, if using. Add the wine, water, and brandy. Season with salt and pepper. Tuck the bouquet garni down the side of the pan.

4 Put a circle of baking parchment slightly larger than the pan on top and pat down. Cover with a tight lid. Bring the stew to a boil, then turn the heat down as low as possible and cook gently for 2 hours. Check after 1 hour to be sure that there is enough liquid—it should come just halfway up the stew. Remove bouquet garni and calf's foot. Serve the stew garnished with chopped fresh parsley.

Braised oxtail with marrow-whipped potato

This hearty, wine-braised oxtail stew finds the perfect partner in smooth, marrow-flavored mashed potatoes.

Serves 4

Ingredients

For the braised oxtail
6 pounds (2.7 kg) oxtail, cut into 1½-inch (4 cm) slices
2 cups (300 g) finely chopped mixture of carrots, celery, and onions (mirepoix)
2 sprigs thyme
3 bay leaves
salt and black pepper
¼ cup (60 ml) olive oil
2 cups (475 ml) red wine
1½ quarts (1.5 l) beef or veal stock

For the whipped potato & bone marrow
2 beef bones, about 4 inches (10 cm) long, split
salt and white pepper
1 pound (450 g) baking potatoes, peeled and chopped
¾ cup (175 ml) heavy cream
¼ stick (30 g) unsalted butter
4 ounces (115 g) bone marrow, strained
4 ounces (115 g) bone marrow, finely diced
2 teaspoons (10 g) chopped fresh chives

Butcher's Tips

Creamy textured, fatty "yellow" beef bone marrow is found in leg bones, and is affectionately known as "prairie butter" because it melts like butter when hot. If your butcher does not have prepared marrow bone, request about 7 pounds (3.25 kg) very fresh bones, cut into slices. These will yield about 8 ounces (225 g) marrow. Rinse the bones before use, then scoop out the marrow from the center, ready to chop or strain. You can also extract the marrow by roasting the bones in a medium-hot oven.

Method

1 Preheat oven to 350°F (180°C). Place the oxtails, vegetables, thyme, and bay leaves in Dutch oven or ovenproof cooking pot and season with salt and pepper. Drizzle with the olive oil. Roast in the oven for 30–40 minutes, or until the oxtails and vegetables are browned. Remove from the oven.

2 Add wine to the pot and simmer over medium heat until reduced by half. Add stock, cover, and braise over low heat for 2 hours, or until the oxtails are tender and falling off the bone.

3 Remove the oxtails and allow to cool slightly. Strain the cooking liquid through a fine-mesh strainer. Return the liquid to the heat and boil until the sauce is thick enough to coat the back of a spoon. Keep warm. Remove the meat from the oxtails, keeping the pieces intact.

4 Make the whipped potato and marrow bones while the oxtail is cooking. Season the split bones with salt and pepper. Place in a baking pan. Roast in the oven for 20–30 minutes, or until the marrow starts to dissolve and the bones are browned.

5 Meanwhile, boil the potatoes in salted water for 15–20 minutes, or until soft when tested with a sharp knife. Take care not to overcook. Drain well and cover to keep warm.

6 In a saucepan, bring the cream to a boil. Mash the potatoes until smooth and season with salt and white pepper. Mix in the hot cream and butter until combined, then stir in the strained bone marrow, diced bone marrow, and chives. Cover to keep warm.

7 To serve, spoon or pipe the whipped potato into the roasted bones. Put a dab of potato onto each plate and sit the bones on top to secure. Serve with the braised oxtails and the sauce prepared from the cooking liquid.

Ossobuco

For this classic Italian stew, ask your butcher for middle-cut ossobuco—meaning "bone-with a hole." The thick-cut veal shanks yield one good-sized piece of meat per serving.

Serves 4

Ingredients

4 sliced veal shanks (ossobuco), weighing about
   4 pounds (1.75 kg) in total
½ cup (60 g) all-purpose flour, to dredge the meat
3 tablespoons (45 ml) olive oil
1 onion, peeled and chopped
2 carrots, peeled and chopped
1 stick celery, chopped
grated zest of 1 lemon
grated zest of 1 orange
1 cup (240 ml) white wine
salt
6 black peppercorns
pinch of saffron powder

Method

1 Dust the veal with flour. Heat 1 tablespoon of the oil in a large skillet or frying pan. Add the ossobuco and fry on each side for 20 seconds, or until lightly browned.

2 In a large Dutch oven or ovenproof pan, heat the remaining oil over medium heat. Add the onion, carrots, and celery and cook gently for 5 minutes to soften. Add the veal in a single layer. Add the lemon and orange zest and cook gently for 5 minutes.

3 Splash on the wine and add just enough water to cover the meat. Add salt to taste, then add the peppercorns and saffron. Cover and cook over very low heat for 1½ hours, until the meat falls away from the bone. Serve with risotto Milanese.

25 Feb, 2020 11:59 AM

Great meat : classic techniques and
award-winning recipes for selecting,
<u>Date Due:</u>   **18 Mar 2020**

<u>To renew your items:</u>

Go online to librariesireland.iii.com
Phone us at (01) 497 3539
Opening Hours: Mon-Thu 10am to 8pm
Fri-Sat 10am to 5pm

Oxtail with tagliatelle

With the wonderful aromas of wine, garlic, and thyme, this no-prep meal is the ideal slow-cooked supper. Serve with fresh tagliatelle pasta, scallions, and Parmesan cheese.

Serves 4

Ingredients

2¼ pounds (1 kg) oxtail
1 onion, peeled and finely chopped
1 clove garlic, peeled
2 sprigs thyme
½ cup (120 ml) red wine
2 cups (475 ml) beef stock
1¼ pounds (500 g) fresh tagliatelle
salt
1 cup (150 g) scallions (spring onions), sliced
1 cup (90 g) grated Parmesan cheese

Method

1 Preheat the oven to 325°F (160°C). Put the oxtail, onion, garlic, and thyme into a Dutch oven or deep roasting pan. Add the wine and stock, plus a little extra water if the meat is not covered. Cover and cook in the oven for 3 hours, or until the meat falls away from the bone.

2 When the meat is cooked, boil the tagliatelle in salted water for 3 minutes, or until just cooked. Drain well. Place the pasta on individual plates and top with the oxtail stew, removing the meat from the bones if preferred. Garnish with scallions (spring onions) and Parmesan cheese.

Oxtail crumble

Precooking oxtail in aromatic vegetables and herbs makes
it extra tasty for this delicate savory crumble and provides
a wonderful base for the sauce.

Serves 4

Ingredients

For the oxtail stock
2 oxtails, cut into pieces
8 shallots, peeled and halved
4 carrots, peeled and chopped
1 head garlic
3 sticks celery, chopped
1 bouquet garni (or 2 bay leaves, 1 teaspoon
   dried thyme, and 1 teaspoon marjoram)
1 cup (240 ml) red wine
salt and black pepper

For the crumble
¼ stick (30 g) butter
8 shallots, peeled and finely chopped
2 cloves garlic, peeled and finely chopped
1 cup (150 g) finely chopped button mushrooms
½ cup (75 g) sliced wild mushrooms

For the topping
5 cups (600 g) all-purpose flour or dry bread crumbs
1 tablespoon (10 g) fresh thyme leaves
salt and black pepper
3½ sticks (400 g) butter, cubed

Method

1 For the stock, soak the oxtail in heavily salted
   water (brine) for 2 hours. Drain, then transfer to a
   large pan. Cover in fresh water, and add the other
   stock ingredients. Bring to a simmer and skim
   off any scum that rises to the surface. Cover and
   simmer for 3 hours, skimming regularly, until the
   meat falls off the bone. Remove the meat.

2 Strain the stock through a fine sieve or muslin.
   In a clean pan, boil the stock with the wine until
   well-reduced and syrupy. Season to taste with salt
   and black pepper.

3 For the crumble, remove the oxtail meat from the
   bones. Melt the butter in a heavy pan. Add the
   shallots, garlic, and mushrooms, and sauté until
   soft. Add the reduced stock and the oxtail chunks.
   Spoon mixture into a deep pie dish. Preheat oven
   to 350°F (180°C).

4 For the crumble topping, mix together the flour
   or bread crumbs, thyme, and seasoning in a bowl,
   then rub in the butter with your fingertips until
   crumbly. Spoon over the meat mixture. Bake for
   40 minutes, or until the topping is golden and the
   meat and vegetables are bubbling and hot.

Braised beef cheeks

Since you cook these cheeks a day in advance, this rich, dark creation makes an effortless dinner that is just the thing for dark winter nights. Serve with mashed baked potatoes.

Serves 8

Ingredients

¼ stick (30 g) butter
2 tablespoons (30 ml) vegetable oil
4 trimmed beef cheeks, seasoned with salt and pepper
2 cloves garlic, peeled and chopped
2 onions, peeled and diced
4 carrots, peeled and chopped
1 leek, washed and sliced
1 stick celery, chopped
6 ripe tomatoes, chopped
3¼ cups (760 ml) red wine
2 quarts (2 l) chicken stock
½ cup (120 ml) veal glace, optional

Chef's Tips

For the accompanying mashed potato, bake 4 large salted potatoes in the oven until tender. Scoop out the flesh and mash with ½ cup (120 ml) milk or light cream until smooth. Transfer the mash to a saucepan, and heat, gradually stirring in 2½ sticks (300 g) cubed butter until melted. Add 1 tablespoon (15 g) finely grated fresh horseradish, and salt and pepper to taste.

Method

1 Melt the butter and oil together in a wide, heavy frying pan over medium heat. Add the cheeks and caramelize all over until they are golden. Transfer the meat to a Dutch oven or baking dish.

2 Add the garlic and vegetables to the frying pan, except for the tomatoes. Cook gently, covered, for 20 minutes. Add the tomatoes and cook for 10 more minutes. Preheat oven to 250°F (120°C). Add the wine, stock, and veal glace, if using, to the vegetables. Bring to a simmer and then add to the cheeks. Cover and braise in the oven for 3 hours, or until the cheeks are soft and tender.

3 Cool the meat, reserving the liquid, then transfer to a deep plate. Cover with baking parchment, then put a chopping board on top and weight down with cans of vegetables or kitchen weights. Refrigerate for 12 hours. Trim off fat from the meat, then cut each cheek in half.

4 In a pan, boil the braising liquid until reduced to about 2 cups (475 ml). To serve, reheat the cheeks in the cooking liquid and serve with mashed potatoes, flavored with fresh horseradish.

Sweet cheat's chili

Chili is the ultimate comfort food, offering tender beef and beans wrapped up in a thick tomato sauce. This quick, child-friendly version is a gentle twist on the original.

Serves 4

Ingredients

1¼ pounds (500 g) ground beef, with 20% fat
salt and black pepper
2 onions, peeled and chopped
2 tablespoons (30 ml) vegetable oil
1 tablespoon (15 ml) tomato paste (purée)
1 cup (200 g) canned kidney beans, drained
2¼ cups (500 ml) beef stock
14½ ounces (415 g) canned baked beans in
   tomato sauce

Chef's Tips

For an "adult" chili, add 1 mild chili (such as Costeño, New Mexico, or Choricero) and 1 hot chili (such as Arbol or Cascabel), both seeded and finely chopped. Add the chiles to the onions in step 1.

Method

1 Season the beef with salt and pepper. Cook in a large pan for 5 minutes, stirring, until evenly browned. Remove and set aside. Heat the oil in the pan and add the onions. Cook gently for 5 minutes, or until softened but not browned.

2 Add the tomato paste, kidney beans, stock, and beef. Simmer for 10 minutes, or until most of the liquid has evaporated. Stir in the baked beans. Simmer for 10 minutes. Serve with potatoes, polenta, tortillas, rice, or tacos.

Beef shank with root vegetables

Fatty bacon peps up this braised stew, imparting essential flavor to the sweet root vegetables. Just make sure your pot is big enough to hold all the ingredients.

Serves 4

Ingredients

3 pounds 5 ounces (1.5 kg) bone-in beef shank (shin of beef), with both ends cut to allow the bone marrow to come out
salt and black pepper
4 ounces (115 g) chopped pancetta or bacon
1 onion, peeled and chopped
1 shallot, peeled and chopped
1 clove garlic, peeled and split
2 cups (475 ml) red wine
1 quart (1 l) beef stock
1 large carrot, peeled and coarsely chopped
1 large parsnip, peeled and coarsely chopped
1 large leek, washed and thickly sliced
1 sweet potato, peeled and coarsely chopped
2 tablespoons (30 g) cornstarch

Method

1 Season the beef with salt and pepper. Heat a large Dutch oven or ovenproof pan over medium heat. Add the beef and brown evenly all over. Remove the beef and set aside.

2 Add the pancetta or bacon to the pan, followed by the onion, shallot, and garlic, then cook over low heat to soften and sweat. Return the beef to the pan and increase the heat slightly. Add the red wine and allow to simmer until reduced by half. Add the stock and remaining vegetables.

3 Preheat the oven to 250°F (120°C). Mix the cornstarch with a little cold water until smooth, then stir into the stew until the gravy thickens. Cover the pan and transfer to the oven. Cook for 4–5 hours, or until the meat falls off the bone.

Chilled Vietnamese braised beef shank

Asian seasonings sit so well with braised beef shank. The flavors intensify as the meat cools, leaving a perfect base for the tangy dressing. Ideal with rice noodles and julienne vegetables.

Serves 4

Ingredients

4 pieces beef shank, about 1.5 inches (4 cm) thick, trimmed and rinsed
1 cup (240 ml) beef or chicken stock
1 cup (about 50 g) chopped fresh cilantro (coriander)
2 star anise
1 stick cinnamon
3 cloves
2 pieces lemon grass, sliced
piece of ginger, ½ inch (1 cm) thick, peeled and sliced
2 cloves garlic, peeled and split
salt and black pepper
2 tablespoons (30 ml) sweet dark soy sauce
¼ cup (60 ml) light soy sauce
¼ cup (60 ml) fish sauce
watercress and watermelon radishes, to serve

For the dressing
juice of 1 lime
¼ cup (60 ml) black Chinese vinegar
3 tablespoons (45 ml) sweet dark soy sauce
3 tablespoons (45 ml) sambal oelek seasoning
2 tablespoons (30 ml) rice vinegar
1 tablespoon (15 ml) sesame oil
¼ teaspoon minced garlic

Method

1 Preheat oven to 350°F (180°C). Place the beef, stock, herbs, spices, and seasonings in a Dutch oven or a large ovenproof pan. Bring to a boil over medium heat, then cover and bake in the oven for 3–4 hours, or until the meat is tender. Remove meat from pot, cool completely, then chill. Strain the cooking liquid and reserve for soup or stock. Slice beef into slices and divide among plates.

2 Mix together the dressing ingredients with a whisk, then toss the beef in the dressing. Serve at room temperature, garnished with thinly sliced watermelon radishes and fresh watercress.

Butcher's Tips

Ask your butcher to trim off the thick cartilage from around the shanks as this is very tough. These offcuts can only be eaten if boiled separately for at least 2 hours, skimming regularly, before chopping and adding to Asian soups, stews, or stir-fry dishes.

Rolled veal roast

For a special occasion, order a boneless veal or beef loin filet from your butcher and stuff with a delicate mixture of ground meat and grated cheese. A real Italian celebration dish.

Serves 4

Ingredients

1¼ pounds (500 g) boneless veal loin or rump
1 tablespoon (15 ml) vegetable oil
¼ cup (50 g) chopped pancetta
¼ onion, peeled and diced
1 carrot, peeled and diced
1 clove garlic, peeled and smashed
3 bay leaves
1 tablespoon (10 g) chopped fresh parsley
½ cup (120 ml) white wine
¼ stick (30 g) butter
2 tablespoons (30 g) all-purpose flour

For the stuffing

1 egg, beaten
1 tablespoon (15 g) grated Parmesan cheese
2 ounces (50 g) finely chopped boiled ham
4 ounces (115 g) ground veal or best-quality sausage
  meat
salt and black pepper
good pinch of grated nutmeg
1 slice soft white bread
1½ tablespoons (20 g) butter, cut into small pieces

Method

1 Mix the stuffing ingredients until well combined.

2 Preheat oven to 425°F (220°C). Use a large, sharp knife to split the meat horizontally in half from one long side to the other, but stopping short of the other edge so the meat can be opened up like a book. Use the knife to stroke the meat along the "spine" to flatten, taking care not to cut through it.

3 Spread the stuffing over the meat, then roll it up and tie loosely with kitchen string. The string will shrink during roasting and, if tied tightly, may cause the stuffing to ooze out. Rub 1 tablespoon oil over the surface and season well.

4 Lay pancetta, onion, carrot, garlic, bay leaves, and parsley in a roasting pan. Place the meat on top. Roast for 45–60 minutes, or until meat registers 145°F (63°C) when tested with a meat thermometer. Rest, covered, for 20 minutes.

5 For the gravy, strain roasting juices, pressing the vegetables to extract the liquid. Deglaze the pan with the wine, stirring all the time. In a saucepan, melt the butter, then whisk in the flour. Whisk in the roasting juices and wine mixture, stirring until thickened. Simmer for 10 minutes. Slice the rested roast veal and serve with the gravy.

Roast chuck eye with potato-mushroom gratin

Also known as a chuck filet or chuck roll, this prime shoulder cut can be dry-roasted to perfection. Here, it's served with a golden baked gratin of cheesy potatoes and earthy mushrooms.

Serves 4

Ingredients

2 pounds (900 g) prime boneless chuck eye roast (eye of the chuck), tied and at room temperature
salt and black pepper
¼ stick (30 g) butter

Butcher's Tips

For a tender chuck-eye roast, ensure that your butcher supplies a prime cut or professionally aged Certified Angus Beef (CAB). If this is not available, select a standard roasting cut, such as a rolled rib roast or loin.

For the potato-mushroom gratin

¼ stick (30 g) butter, cubed, plus extra for greasing
4 Yukon Gold potatoes, peeled
1 cup (150 g) cooked mixed mushrooms, wild or cultivated
1 cup (about 115 g) grated Gruyère cheese
½ cup (45 g) grated Parmesan cheese
1½ cups (350 ml) heavy cream
1 teaspoon (5 g) sea salt
½ teaspoon ground white pepper
¼ teaspoon ground nutmeg

Method

1 Preheat oven to 275°F (140°C). Season the beef with salt and pepper. Melt the butter over medium heat in a large sauté pan. Add the meat and brown on all sides.

2 Transfer the meat to a baking rack over a baking pan. Bake in the center of the oven until the internal temperature of the meat reaches at least 125°F (52°C) in the center—this will leave the meat rare, but it will continue to cook a little when resting and reheated for serving. You can also cook meat to an internal temperature of 140°F (60°C) for medium-done. Allow the meat to rest, covered, for 15 minutes, then cut into thin slices across the grain.

3 For the gratin, grease a large, deep baking dish with butter. Increase the oven heat to 350°F (180°C). Cut the potatoes into very thin slices with a sharp knife or mandolin slicer. Place a single layer of sliced potatoes in the dish, then add a layer of cooked mushrooms and grated cheese. Layer the remaining potatoes, mushrooms, and grated cheese in the same way.

4 In a bowl, season the cream with salt, pepper, and nutmeg. Slowly pour the mixture over the potatoes and press on the mixture to ensure potatoes are covered. Top with small cubes of butter. Bake for 40–50 minutes, or until the potatoes are tender when tested with a sharp knife and the surface of the gratin is browned.

5 About 5 minutes before the gratin is finished cooking, return the sliced beef to the oven, covered, to reheat. When piping hot, serve the beef with the gratin.

English roast beef & Yorkshire pudding

Nothing could be more British than roast beef and Yorkshire pudding served for Sunday lunch. This classic meal needs only the simplest ingredients and the best beef you can buy.

Serves 6–8

Ingredients

6 pounds (2.7 kg) standing rib roast (bone-in rib of beef), 3–4 ribs in total
salt and black pepper
1 tablespoon (10 g) dry mustard powder
2 small onions, peeled and halved
3 carrots, peeled and halved
1 tablespoon (15 g) all-purpose flour

For the Yorkshire puddings (see Chef's Tips)
2 cups (475 ml) all-purpose flour
2 cups (475 ml) milk
2 cups (475 ml) beaten eggs

Chef's Tips

For Yorkshire puddings, mix batter until smooth, then rest it for 1 hour. While the beef rests, pour the beef fat into a baking pan or muffin pans—add extra oil if there is not much. Bake the pan at 450°F (230°C) until smoking. Add the batter and bake for 10–20 minutes to set. Open the oven briefly to release steam. Bake again for 10 minutes at 350°F (180°C).

Method

1 Preheat oven as high as it will go—450°F (230°C). Rub the mustard, salt, and pepper over the meat. Put the onions and carrots into a roasting pan with the meat. Roast for 20 minutes.

2 Turn oven down to 350°F (180°C). Roast beef for 15 minutes per pound (450 g) for rare; add 15 minutes to the total cooking time for medium-rare, 25 minutes for medium, and 35 minutes for well-done. Check the beef regularly, basting at the same time. If the meat feels soft, it is rare. Medium-rare meat feels fairly firm when prodded in the center. Allow cooked meat to rest, covered, for 30–60 minutes. Pour the cooking juices into a bowl, allow to settle, then skim off most of the fat and reserve.

3 For the gravy, pour 1 tablespoon reserved fat back into roasting pan with carrots and onions. Stir in 1 tablespoon flour. Heat gently, stirring, for 1 minute, then add reserved pan juices and juices from rested meat. Cook, stirring, until smooth and thick. Serve with the beef, Yorkshire puddings (see left), and steamed seasonal vegetables.

Roast beef on hot dripping toast

The best and tastiest way to use up leftover roast beef, serve this tasty combo as a first course.

Serves 4

Ingredients

1 cup (190 g) cold beef dripping, reserved from roast
    beef or sourced fresh from your butcher
4 thick slices fresh bread, crusts removed if preferred
leftover roast beef, about 4 ounces (115 g) per person
handful of flat-leaf parsley leaves or salad greens
2 tablespoons (20 g) capers
4 tablespoons (60 ml) horseradish sauce
sea salt

Method

1 Preheat oven to 350°F (180°C). To make your
   own dripping, cut off all the excess fat from
   your cooked roast beef and put into a baking
   pan. Bake for 30 minutes, or until the fat melts.

2 In a large frying pan, fry the bread in the dripping
   over low heat until it is crispy and golden on
   the outside, but still soft in the center. Heat the
   beef for 1–2 minutes in the oven, and warm the
   horseradish sauce in a small pan or microwave
   for a few seconds.

3 Serve the toast with the beef on top, accompanied
   by the parsley or salad leaves, capers, and
   horseradish sauce. Season with salt.

chapter 2

Pork

Pork is one of the most loved meats because it is such good value. Pork suits a range of welcoming dishes and offers infinite options to preserve and cure its tender meat. For every day or a special celebration, there is a cut for every occasion.

From Farm to Table

Pigs have been farmed for hundreds of years—longer than beef or lamb. They need little space for grazing and are legendary omnivores, happy to make a meal of just about anything. This mix of foods contributes to great-tasting meat.

A HAPPY PIG IS A TASTY PIG

Intensively reared pigs live inside specially designed pig warehouses—known as barns, piggeries, pig sheds, or hog lots—with little space to move and hardly any freedom to exhibit natural behavior. Many consumers now prefer free-range pigs that are raised on pasture, both for the higher welfare standards and the vastly superior flavor of the meat. However, only about 4% of pigs spend their entire life outdoors; most are brought to inside to be fattened for market.

Animals are only as good as the food they eat, so a varied diet of grain and vegetables means tastier pork on the plate. Intensively farmed pigs eat a mix of nutritional grains and extra protein, including soy, meat, and bone meal which can produce meat with a bland taste. But meat from a free-range pig has lashings of intense flavor. The flesh is a strong shade of pink, with a marbling of creamy white fat to keep it moist. The difference in these two farming methods is reflected in the pork's cost and quality—the more you pay, the better the quality. Pork does not have tough connective tissue and so doesn't need to hang and mature after slaughter.

FAVORITE BREEDS

There are hundreds of hog and pig breeds that can be raised for meat. The massive American Yorkshire, also known as the English Large White, has been a staple for nearly 200 years. Spotted Berkshires are slightly smaller, but equally hardy and fertile. Chesters and Hampshires originate in the U.S., but are descended from pure English stock. Durocs are the most efficient breed for converting feed to final weight and are popular with farmers who rear intensively. Some farmers raise rare breeds for showing, such as Gloucester Old Spot and Tamworth, but there are so few of these animals that they are rarely slaughtered for meat. Some say that breeds taste different, but how the animals are raised is more important to how they taste. Different breeds do have different body shapes: pigs with extra lean hindquarters are not great for hams, and larger breeds with more fat are suited to bacon.

Bacon and Ham

Our love of salty cured pork—eaten as ham or bacon—is so intense that this is where much reared pig meat ends up. Pigs bred for curing are larger, older, and fatter than those animals used for pork cuts, but essentially the cuts are similar and recognizable. For example, Canadian (back) bacon is made from the same area as loin chops, and regular (streaky or side) bacon is taken from the richly fatted belly of the animal. Ham is the name we give to the specially cured meat from the rear leg, whereas a ham hock is cut from the smaller front leg.

Nature's Tractor

Pigs forage on leaves, grass, roots, fruits, flowers, brambles, acorns, and even earthworms. They are such effective diggers that when foraging on land previously planted with vegetables, they leave it dug and manured for the next crop. "Pig tractoring," is used by small-scale farmers to bring overgrown land into fresh production.

CURING

To produce bacon (or ham) from pork, the meat is first cured in salt—either with a wet brine that is soaked or injected into the meat, or as a dry salt rub (dry curing). The brine can be flavored with herbs or spices, or the preserving liquor is composed of something other than salted water—beer, cider, or apple juice. In both brine and dry-cured meats, natural sugars can be added for extra sweetness, as in traditional maple-cured bacon. Smoking the cured bacon or ham adds additional flavor, and also improves its keeping qualities.

BACON

Mainstream customer preference means that it's the norm for American bacon to come ready smoked, while in the U.K. and Ireland both smoked and unsmoked (or green bacon) are equally common. Likewise, American bacon is almost always cut from the belly—sold as side bacon in the U.K. In contrast, Canadian (back) bacon comes from the center-back of the pig, rather than the belly. Because it contains less fat, Canadian bacon offers more meat per serving, but is not as crispy when cooked. Canadian bacon maybe sold under the name back bacon, Irish bacon, or short-back bacon. Traditional Canadian bacon (sometimes known as "peameal" bacon) is cured in a light brine, left unsmoked, and may be finished with cornmeal coating. Italian pork is also cured unsmoked. Known as pancetta, it looks just like regular bacon when sliced or cubed. The French use cubes of very fatty bacon—known as lardons—in many classic recipes using meat, cheese, or eggs.

Popular across the Americas, Asia, and Europe, hams are made from the pig's hindquarters. They are cured in salt, sugar, and often nitrate preservatives, using a wet or dry preserving process. Serrano and prosciutto ham are renowned hams from Spain and Italy—the outside of the ham is rubbed with salt and spices, then left to dry before it is sliced into wafer-thin layers. The curing process can take a few months or several years! These hams are classified according to the breed and feed of the pig, and the time allocated to curing.

Sausage

Preserving pork trimmings is the realm of the sausage-maker. The thrifty housewife's need to use all of the pig has resulted in the creation of hundreds of seasoned delicacies—raw, cured, dried, smoked, pickled, or aged to perfection. The wide range of sausages is reflected in their culinary use: they can be employed as a main recipe ingredient, as a subtle flavoring, or as a decorative garnish. A skilled butcher is trained in the fine art of raw sausage-making and can tell you exactly what is inside the sausage casing.

Prized Jamón Ibérico pigs forage naturally on herbs, grasses, and acorns—*bellota*—in Spain's ancient oak woodland pastures.

Grill & Broil

Sizzling under the broiler; smoking on the barbecue—grilled or broiled pork is the easiest option for the meat-loving cook. The tender meat welcomes other flavors, whether it be a Cajun spice rub or Asian marinade.

Pork for the grill or broiler needs fat to keep it moist. Do not season too far in advance or the juices will be drawn out, but do make sure you add plenty of herbs and spices to jazz up the meat. A key point is the thickness of the cut: thin chops dry out easily so they are cooked quicker, while thicker chops need time to cook through and so are grilled more slowly. Lean cuts also benefit from quick cooking to prevent them from drying. Pork is never served rare and the resting time is essential: Cover the cooked meat for 10 minutes to let the fibers relax and allow the appetizing flavors to develop.

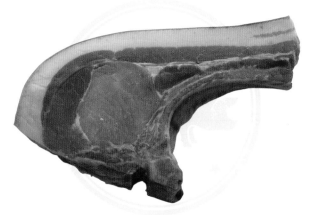

Rib Chop

Near the front of the pig, the rib chop is a thick bone-in cut. The meat is lean, but surrounded on one side by a layer of fat that keeps it moist. Score into the fat before cooking to prevent the chop from buckling on the grill. These chops are also available boned and the thinner end is sold as spareribs.

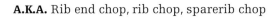

A.K.A. Rib end chop, rib chop, sparerib chop

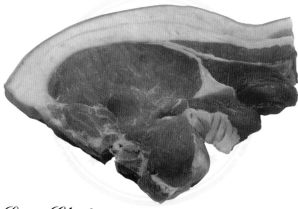

Loin Chop

The pork equivalent of a T-bone steak, this cut comes with a layer of juicy fat and a rind that can be turned into an individual portion of crisp crackling. Before cooking, simply score through the rind, rub with salt and oil, then use tongs to hold rind-side down on a hot griddle until crisp. Your butcher can also slice across the eye of the loin to make medallions.

Tenderloin Scallop

From the filet that runs along the top of the pig, slices of tenderloin can be flattened to form scallops or left round for medallions or noisettes. Since this is a delicate cut and lacks the fuller flavor of harder-working muscles, it is vital that you don't overcook it as the fine fibers can turn rubbery. Cook with herbs, spices, onions, or garlic to boost the flavor of the finished dish.

A.K.A. Fillet, center loin, tenderloin escalope

Other Pork Grill Cuts

- **SIRLOIN END** This smaller pork cut comes from the rear of the loin. It is bursting with flavor and doesn't need long cooking. Ask the butcher to leave a layer of fat around the "eye" to keep it moist. Also known as rump or chump chops.

- **FORE LOIN** This specialty cut from the front of the loin provides standard cutlets.

- **MIDDLE LOIN** Another excellent source of grilling and broiling chops.

- **SHOULDER (COLLAR) CHOPS** One of the most economical cuts, these small chops are fatty and do not dry out during cooking.

- **BELLY SLICES, SLAB BACON, AND RASHERS** There is so much fat on the belly that it never dries out when grilled.

Roast, Braise & Stew

Pork is a very accommodating and versatile—the cuts that are good for roasting are generally good for braising, pot-roasting, and stewing as well.

The meat from the forequarters of the pig tends to have the most well-developed muscles, marbled with juicy fat and distinctive qualities of taste. The meat in the center of the animal is the most tender, while the cuts at the rear are the leanest and most prone to drying out during cooking.

Cubed Meat

Pork is often sold pre-cut for stewing, pot-roasting, and braising. This provides value and convenience for the cook, but it's best to check that the meat comes from the same cut as they all have different cooking needs. Buy from a good butcher to ensure you're not being sold mixed random trimmings.

Spareribs

A single bone-in sparerib is a winning cut for braising: After cooking the bones just slip out of the soft meat. This is an economical but meaty cut, offering maximum flavor. Ask your butcher to trim it for you.

A.K.A. Rack of ribs, rib rack

Shoulder (cubed)

Meat from the shoulder is sold ready-cubed for stewing and braising. It offers masses of flavor but can be a little on the tough side. Best-suited to very slow cooking or pot-roasting in stock, wine, or cider.

A.K.A. Pork stewing meat

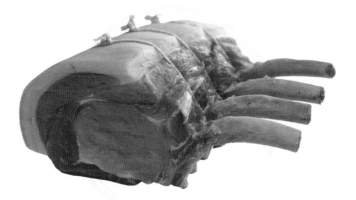

Rack of Pork

An impressive roast with moist, tender meat and crispy crackling. A skilled butcher can also prepare it boned-out, ready for stuffing and rolling, although cooking on the bone delivers more flavor.

A.K.A. Frenched loin, center rib roast

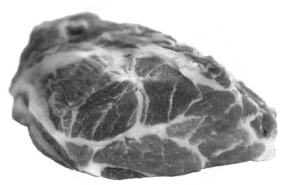

Shoulder

Rich in creamy fat and connective tissue, this cut is best boned, rolled, then cooked slowly to allow it to tenderize. The cooked meat is fall-apart-tender when teased away with a fork.

A.K.A. Butt, Boston butt, blade, hand of pork

Belly

This is the cut employed to make bacon. But when left unsalted and in one piece, belly pork is a super slow-cooked cut. After several hours in a low, low oven, the meat simply flakes apart. The crackling crisps up well for a hungry chow down.

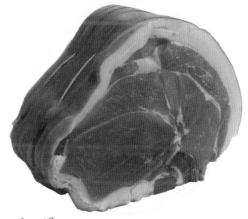

Fresh Ham

The leanest, and best value, roast. Ideal as a standard roast or as super-slow "pulled pork." For supreme crackling, apply a salt rub to the scored skin before cooking. Ask the butcher to tie the roast.

A.K.A. Boneless ham roast, leg

Snout to Tail

There is an old saying that with pigs you can eat everything but the squeak. Here's how to get the best culinary results from the less-known parts of the animal.

The golden rule when buying organ meats (offal) is "fresher the better." If these cuts are not available at your local store, you'll need to find a butcher who knows where his or her meat comes from and can take responsibility for its quality.

Head

The head is the key ingredient in the age-old delicacy, head cheese. This is a seasoned terrine made with the boiled head meat and/or trotters. Gelatin from the same animal is traditionally used to set the meat. There are many regional variations, and the dish is prized for its local variations.

A.K.A. Hogs head cheese, souce, brawn

Cheek

This nugget of super tasty muscle was once only used for ground meat or stock. Now it is sought out by the best chefs for slow-braised dishes. There are only two small morsels of cheek per pig—enough for one person—so they are rare and highly prized.

Liver

A just-cooked slice of liver from the freshest outdoor-reared pig comes silky smooth, yielding, and not bitter at all. It can benefit from soaking in milk for an hour before cooking, but this is unnecessary if your butcher is confident of its freshness.

Heart

The heart is the hardest-working and toughest muscle. Best cooked very, very slowly: First simmer gently in water until tender, then pack with a traditional sage-and-onion stuffing. Braise in a slow-cooker and keep moist with a little stock. Ask your butcher to trim the heart ready for cooking.

Tail

Score deeply across the skin, rub with salt, honey, and a little oil, then roast quickly in a hot oven to make the best home-made bar snacks. Or, braise very slowly with root vegetables, then drain, cool, and chill. Cover the tails with a light crumb coating, then roast in butter until crispy.

Kidney

Pigs' kidneys are ideal for mixing with beef in an English steak and kidney pie. Or, blend with bacon for superb patties. On their own, "deviled" kidneys are prepared by cooking chopped kidneys in a spiced, jelly-wine sauce. Ask your butcher *exactly* what the pig ate when reared as this will affect the taste of the kidney. Only buy fresh kidneys whole and with the surrounding fat.

Trotter

Always boiled before use, trotters can be boned by your butcher ready for stuffing and roasting. Alternatively, they can be braised with stock, vegetables, and herbs. After three hours slow-cooking, you will be able to remove the tender meat from the bone and strain off the cooking liquor. The meat is ideal for pâtés, while the gelatinous liquor can be bottled and refrigerated as fresh pork stock.

Pretzel-crusted pork chop Viennoise

Crumbed pork chops are a twist on the classic Austrian veal dish—Wiener Schnitzel—and are served with the traditional accompaniments of hard-boiled eggs, capers, onion, and lemon.

Serves 4

Ingredients

4 boneless pork chops, each weighing about
    10 ounces (280 g), pounded with a meat mallet
salt and black pepper
2 ounces (50 g) all-purpose flour, sifted
4 eggs, beaten
2 cups (180 g) pretzel bread crumbs
½ stick (60 g) butter

For the garnish
1½ sticks (170 g) butter
juice of 1 lemon
salt and black pepper
4 hard-boiled eggs, whites and yolks
    chopped separately
1 small red onion, peeled and minced
½ cup (60 g) capers
2 tablespoons (20 g) chopped fresh parsley
¼ cup (60 ml) mustard oil or olive oil
1 lemon, cut into 4 wedges

Method

1 Preheat the oven to 350°F (180°C). Season the pork chops with salt and pepper. Put the flour into a shallow bowl, the eggs into a second bowl, and the pretzel bread crumbs into a third. Coat the pork chops with the flour, then dip in the eggs, and then coat evenly in the bread crumbs.

2 Melt the butter in a skillet and sauté the coated pork chops until golden brown on both sides. Transfer to a roasting pan and bake in the oven for 4–5 minutes, or until the meat is cooked to medium-done. Remove and keep warm.

3 For the garnish, heat the butter in a pan over low heat until it turns brown and the milk solids have separated, or until you hear no more crackling from the butter. Add the lemon juice, then season with salt and pepper.

4 Serve the pork chops with the eggs, onion, capers, and parsley. Drizzle with the browned butter and mustard or olive oil, and top with lemon wedges.

Pork chops with apple & cider

Pork chops are cooked in a rich, creamy, apple-brandy sauce.
This recipe originates in the Normandy region of France, which
is famed for its apples, Calvados brandy, and heavy cream.

Serves 4

Ingredients

¼ stick (30 g) butter
3 large cooking apples, peeled, cored, and cut into
  quarters
1 small leek, trimmed and chopped
salt and black pepper
3 tablespoons (45 ml) hard apple cider or Calvados
  brandy
2 cups (475 ml) apple juice
4 French-trimmed pork chops
1 cup (240 ml) heavy cream

Chef's Tips

The sweetness of the apple brandy and apple
juice counteracts the tart flavor of the apples.
If you don't have apple brandy, use ordinary brandy
instead—the recipe will work just as well.

Method

1 Melt 1 tablespoon of the butter in a large skillet
   over medium heat. Add the apples and chopped
   leek. Cook, stirring occasionally, until slightly
   browned. Season to taste with salt and pepper,
   then pour in the apple cider. Let this bubble
   up and cook until almost all the liquid has
   evaporated. Pour in the apple juice and bring
   to a boil. Cook for 3 minutes, then set aside.

2 Heat another skillet over medium heat and add
   the remaining butter. Add the chops and cook for
   2 minutes on each side, or until browned. Remove
   the pan from the heat.

3 Return the pan of sauce to the heat and bring
   back to a simmering point. Add the seared pork
   chops and cook for another 3 minutes, or until the
   meat is cooked through. Reduce the heat, pour in
   the cream, and stir until well combined. Season to
   taste, then serve the chops with the sauce poured
   over the chops.

Classic Neapolitan ragu with involtini

Every Italian mama has her own recipe for this classic Sunday meal. This one is from Luigi Lino's mother, Mathilde, who usually puts the involtini on a platter, and serves the sauce with pasta separately.

Serves 4

Ingredients

For the sauce
1 large onion, peeled and diced
3 cloves garlic, peeled and minced
2 sticks celery, diced
1 large carrot, peeled and diced
2 tablespoons (30 ml) olive oil
6 pork spareribs, cut in half
7 ounces (200 g) ground (minced) beef
¾ cup (175 ml) red wine
1¼ pounds (500 g) canned plum tomatoes

For the involtini
2 beef sirloin (rump) steaks, 8 ounces (225 g) each
2 tablespoons (30 g) grated Parmesan cheese
2 tablespoons (20 g) chopped fresh parsley
1 clove garlic, peeled and thinly sliced

Butcher's Tips

The combination of ground meat and steak can provide two separate meals from one dish. The sauce makes a great partner for pasta or polenta.

Method

1 In a large, heavy pan, fry the onion, garlic, celery, and carrot in the oil until soft. Add the pork ribs and ground beef and cook for 2 minutes. Stir in the red wine and plum tomatoes. Fill the empty tomato can with water and pour into the pan.

2 While the sauce comes to a boil, make the involtini. Pound the beef steaks with a meat mallet until thin. Lay 1 steak on a board and sprinkle over half the Parmesan cheese, parsley, and garlic. Roll it up like a cigar and tie with kitchen string. Repeat with the second piece of beef. Carefully lower the involtini into the sauce, cover, and simmer very gently for 2 hours.

3 Carefully lift the involtini from the sauce, remove the string, and cut into slices before serving with the meat sauce and pasta.

Braised pork cheek with green lentils & root vegetables

Tender meat is coated with crisp bread crumbs, then served on a bed of tasty vegetables. Cutting the vegetables into tiny dice is time-consuming, but the final result is well worth it.

Serves 4

Ingredients

4 pork cheeks, weighing about 5 ounces (140 g) each
salt and black pepper
½ stick (60 g) butter
1 small onion, peeled and finely diced
1 carrot, peeled and finely diced
1 stick celery, finely diced
1 head garlic, split into cloves and peeled
2 tablespoons (30 ml) tomato paste (purée)
2 teaspoons (10 g) chopped fresh tarragon
2 cups (475 ml) brown chicken or veal stock
all-purpose flour, for dusting
3 eggs, beaten
3 tablespoons (45 ml) whole grain mustard
1 cup (about 100 g) panko bread crumbs

For the green lentils & root vegetables
3 tablespoons (45 g) butter
¼ cup (about 35 g) peeled and finely diced carrot
¼ cup (about 35 g) peeled and finely diced celery root (celeriac)
¼ cup (about 35 g) peeled and finely diced turnip
¼ cup (about 35 g) peeled and finely diced leek
½ teaspoon chopped fresh tarragon
1½ cups (300 g) green French lentils, cooked

To garnish
fresh watercress

Method

1 Season the cheeks with salt and pepper. Heat 2 tablespoons of the butter in a skillet, then brown the cheeks on both sides. Add the vegetables and cook gently for 5–6 minutes. Add the garlic, tomato paste, and tarragon. Cook for 2 minutes. Stir in the stock. Transfer to a pot with a tight-fitting lid. Simmer for 2–3 hours, or until cheeks are tender.

2 Remove the cheeks and leave to cool. Strain the sauce through a fine sieve, reserving the liquid and discarding the vegetables.

3 For the lentils and root vegetables, gently heat 2 tablespoons of the butter in a pan. Add the finely diced vegetables. Season with salt and pepper, and cook over low heat for 6–8 minutes, or until lightly browned, but still firm. Add the tarragon, lentils, and 1 cup of reserved pork cooking liquid. Simmer until the liquid is reduced by half, then stir in the remaining 1 tablespoon butter.

4 Slice the cooled pork cheeks in half horizontally. Put the flour in a shallow bowl, the eggs and mustard in a second bowl, and the bread crumbs into a third. Coat the pork cheeks with flour, dip in the eggs, and then coat in the bread crumbs.

5 Melt remaining 2 tablespoons butter in a skillet. Add the cheeks and cook over medium heat until they are browned evenly on both sides. Remove from the heat and leave to rest in a warm place while you reheat the green lentils and root vegetables. Spoon the lentils into a shallow bowl and top with the pork. Garnish with watercress.

Chef's Tips

The French terms "mirepoix" and "brunoise" are sometimes used to refer to a mixture of finely diced vegetables: Simply cut the vegetables into matchsticks, then cut across into tiny cubes.

Honey-glazed pig's tails

Here's the proof that you can eat every bit of the pig except the squeak. Braising, then roasting the tails with lashings of honey cooks these morsels to absolute perfection.

Serves 4

Ingredients

8 pig's tails, hairs removed
1–2 inches (2.5–5 cm) piece fresh ginger
3 cloves garlic
olive oil
salt
¾ cup (about 100 ml) clear honey

Butcher's Tips

The very end of a pig's tail is just cartilage and so there's no good in eating that! But a few inches down the tail, it becomes beautifully fleshy and fatty. When you go to the butcher, ask him to trim off the tail's tip. Ask for good long tails—the tightly curled sort won't be as good—with the skin ready-scored.

Method

1 Place the pig's tails in a large saucepan and cover with water. Peel and roughly chop the ginger and garlic, and add to the water. Cover the pan and bring to a boil. Reduce the heat and simmer for 2–3 hours, or until the tails are tender. Remove the tails from the cooking liquid and leave until cool enough to handle. You can discard the cooking liquid or strain and use as a stock.

2 Preheat the oven to 400°F (200°C). Place the tails in a roasting pan and drizzle with olive oil. Use your hands to rub in the oil until the tails are well coated. Sprinkle each tail with salt to taste, then spread over a liberal amount of honey until completely coated. Roast for 45–50 minutes, or until the skin is dark and sticky.

Sticky spareribs

Juicy spareribs with a lip-smacking glaze make perfect finger food—just be sure to supply your guests with finger bowls and plenty of napkins.

Serves 4

Ingredients

4½ pounds (2 kg) pork ribs, cut into portions of 3–4 ribs each
salt
½ cup (60 ml) clear honey
5 tablespoons (70 ml) whole grain mustard
1 teaspoon (5 g) cayenne pepper
1 teaspoon (5 g) Chinese five-spice powder

Butcher's Tips

Spareribs are also delicious cooked on the barbecue as the smoke adds to the flavor of the meat. Grill slowly over medium heat until the meat is just ready to fall off the bone.

Method

1 Place the ribs in a large pot of cold, salted water and bring to a boil. Simmer for 45 minutes to tenderize the meat. Drain and leave to cool.

2 Mix the remaining ingredients in a large bowl, then rub over the cold ribs with your hands, making sure they are completely covered with the marinade. Cover with plastic food wrap and chill for 8 hours, or overnight, to marinate.

3 When you are ready to cook the ribs, preheat the oven to 275°F (140°C). Place the ribs in a large baking pan with the marinade. Roast for 1½ hours, basting occasionally, until tender.

Sausages

Everyone loves sausages! You can buy good sausages from your butcher, but if you make your own you can try different flavors.

Who can resist a crisp, crackling sausage sizzling in the smoking hot skillet or over the barbecue? Whether served on its own with a buttery mound of mashed potatoes or added to a traditional French cassoulet, the sausage is the meat-lover's ultimate homage to frugal country living because it is made from the unwanted trimmings from the butchered animal.

FLAVOR AND TEXTURE PALETTE

With homemade sausages, you know exactly what is going into them from the start—unlike some mass-produced sausages. Sausages need a basic combination of ground lean and fatty meat, plus seasonings. You can add all manner of flavorings to the meat: a little red wine, garlic, and fresh thyme added to ground pork produces a French Toulouse sausage, while a spoonful of harissa paste with ground lamb creates a North African merguez sausage.

A good sausage is a combination of different meat proteins. As the sausage cooks, the combined meat firms up, traps fat, juices, and also the added flavorings. Fat and salt give sausages an advantage over whole meat cuts by retaining more moisture, flavor, and by imparting juiciness. Salt is also responsible for a moist interior sausage texture because it helps dissolve the tough muscle fibers in the ground meat, allowing it to stay soft and prevent shrinkage.

HEALTH AND HYGIENE

Because raw sausage meat has an open surface area on which bacteria can multiply, it is vital that your hands, utensils, and all sausage-making equipment is kept clinically clean, and that your raw sausage ingredients are kept chilled at all times, even between recipe steps. Never taste raw meat to check the seasoning; instead, fry a small patty of sausage meat in a skillet, then taste.

Sausage skins

Sausage skins—or casings—come in two types. Artificial skins are made with collagen—an edible animal protein (usually cattle) that is processed to produce thin, smooth casings into which the sausage meat is forced. Alternatively, you can use natural casings which are animal intestines (hog or sheep) that have been cleaned. Natural casings are soaked in water before use to remove the excess salt and help the casings slide over the stuffing. Natural casings are the preferred choice for epicures who are going to the trouble of making their own sausages and want to make them as authentic as possible.

COOKING FOR SUCCESS

The best way to cook sausages is to grill them over the barbecue on an oiled rack or fry gently in a skillet with a little vegetable oil. Roll or turn the sausages regularly using a turner or metal spatula for even browning. As the sausages darken in color, some areas will caramelize to a sticky deliciousness but should not be allowed to burn. Do not be tempted to rush the cooking or you risk splitting the thin sausage skins. Fast cooking will also cause your sausages to burn on the outside and stay raw at the center.

At the other extreme, if you cook sausages too slowly over low heat, they will be pale, dry, and wrinkled, without having sufficient heat to brown the outside. Never, ever be tempted to prick the skin while cooking, since the sealed casings serve to retain all the flavors and fat. Always cook sausages until they are hot throughout, above 155°F (68°C) when tested with a meat thermometer.

Herbed pork patties

Tender organ meats grind up beautifully for these wrapped, herbed patties. To make the mashed potatoes even richer, stir in an egg yolk once the butter and cream have been incorporated.

Serves 2

Ingredients

For the patties
4 ounces (125 g) pig's liver, ground (minced)
4 ounces (125 g) pig's heart, ground (minced)
4½ ounces (135 g) ground (minced) pork
1 onion, peeled and finely chopped
1 sprig rosemary, finely chopped
1 sprig thyme, finely chopped
salt and black pepper
2 cups (250 g) fresh bread crumbs
4 ounces (125 g) caul fat
1 cup (240 ml) beef or pork stock

For the mashed potatoes
11 ounces (300 g) red-skinned potatoes, peeled
  and cut into even-sized pieces
½ cup (120 ml) heavy cream
½ stick (60 g) butter
salt and black pepper

Method

1 Preheat the oven to 325°F (160°C). Mix together the liver, heart, and ground pork, then stir in the onion, herbs, seasoning, and enough bread crumbs to form a firm mixture.

2 Soak the caul fat in cold water until it is soft and stretchy. Drain well, then lay on a clean work surface. Place heaped tablespoonfuls of the patty mixture on the fat, trim the fat with a sharp knife around the ball of meat, then carefully wrap the patties in their casing. Repeat until all the mix has been used.

3 Place the patties in an ovenproof dish and pour over the beef stock. Bake for 30–40 minutes, or until cooked through.

4 While the patties are braising, prepare the potatoes. Boil the potatoes in salted water until tender. Drain. Put the cream and butter into a small saucepan and bring to a boil. Mash the potatoes until smooth, then stir in the hot cream and butter and season with salt and pepper. Serve with the patties.

Pork belly braised in cider

Pork belly really benefits from a slow-cooking technique, such as the braising method used here. You can replace the cider with apple juice, but the sauce will be sweeter.

Serves 4

Ingredients

2 onions, peeled and coarsely chopped
2 carrots, peeled and coarsely chopped
2 parsnips, peeled and coarsely chopped
1 clove garlic, peeled and finely chopped
1 tablespoon (10 g) chopped fresh sage
2¼ pounds (1 kg) pork belly, with skin scored
1 quart (1 l) hard cider (dry cider)
salt and black pepper
2–3 tablespoons (30–45 g) cornstarch (cornflour)

To serve
creamy mashed potatoes and green cabbage

Butcher's Tips

The great thing about this dish is that you can do all the main cooking in advance. Braise the pork belly at the lower temperature for 3 hours the day before you plan to serve. Then simply finish off at a high heat the next day, and make the sauce just before you're ready to serve.

Method

1 Preheat the oven to 250°F (130°C). Mix the vegetables, garlic, and sage in a bowl. Place the pork, skin-side up, in a Dutch oven or roasting pan. Add the vegetables and enough cider to cover meat. Season well. Cover and braise for 3 hours.

2 Increase the heat to 400°F (200°C). Strain the cooking liquid from the meat into a pitcher. Transfer the vegetables to a separate dish, cover, and keep warm. Return the pork to the roasting pan and sprinkle the skin with salt. Bake for 30 minutes, or until the pork skin is crisp.

3 Meanwhile, make the sauce. Measure the cooking liquid: for each cup of liquid you will need 1 tablespoon (15 g) cornstarch. Place the cornstarch in a bowl and add the same number of tablespoons of cooking liquid. Stir to form a thin paste. Pour remaining cooking liquid into a pan. Whisk in the cornstarch mixture and gently bring to a boil, stirring. Simmer, stirring, to thicken.

4 Remove the pork from the oven and leave to rest, covered, in a warm place for 30 minutes. Serve the meat with the sauce and reserved vegetables, accompanied by potatoes and cabbage.

Boiled ham hock & pease pudding

This hearty dish makes an ideal lunchtime meal. The split yellow peas, traditionally known as "pease pudding," are cooked in the ham stock and absorb all its wonderful flavors.

Serves 2

Ingredients

For the ham
1 smoked ham hock
1 small onion, peeled and halved
2 cloves garlic, peeled
½ leek
2 sticks celery
4 carrots, peeled
sprig of thyme
sprig of rosemary
sprig of parsley
10 black peppercorns
2 cloves
1 star anise
1 teaspoon (5 g) caraway seeds
1 cup (225 g) split yellow peas, soaked overnight
flat-leaf parsley, to garnish

To serve
buttered bread

Method

1 Place the ham hock in a large pan, cover with cold water and bring to a boil. As soon as it has boiled, pour off the water and replace with fresh water. Add the onion, garlic, leek, celery, 1 of the carrots, herbs, and spices to the pan. Bring to a boil, then simmer for 3 hours, adding water occasionally to keep the ingredients covered.

2 When the hock is cooked, leave to cool. Place the yellow peas in a pan and pour over 2 cups (475 ml) cooking liquid from the ham. Simmer for 45 minutes, or until peas are mushy but not reduced to a purée. Slice the remaining 3 carrots lengthwise into thick batons and boil in the remaining ham stock until tender. Drain.

3 To serve, remove the skin from the ham and discard, then flake the meat—it should just fall away from the bone. Place a spoonful of the peas in the center of each plate. Add the carrots and ham, then garnish with parsley and serve with buttered bread.

Classic sausages with bubble & squeak

Bubble and squeak is a great way to use up leftover vegetables—traditionally potatoes and cabbage. Which one is the bubble, and which one the squeak, we may never know.

Serves 4

Ingredients

8 breakfast sausages of your choice
2 onions, peeled and thinly sliced
1 tablespoon (15 g) all-purpose (plain) flour
2 cups (475 ml) beef stock
2 tablespoons (30 ml) vegetable oil or butter
2 pounds (900 g) cooked, peeled potatoes, cubed
½ green cabbage, cooked and shredded
salt and black pepper

Chef's Tips

Bubble and squeak is usually made with leftover potatoes and vegetables, most commonly cabbage and carrots from a traditional Sunday lunch. The potatoes can be roasted, mashed, or boiled.

Method

1 Preheat the oven to 350°F (180°C). Place the sausages in a roasting pan and roast for 30 minutes, turning occasionally, until golden brown. Remove from the oven and keep warm.

2 Pour the sausage fat from the roasting pan into a saucepan. Add the onions, and cook over high heat, stirring constantly, for about 5 minutes, or until golden brown. Stir in the flour, then add the stock and simmer, stirring, until the gravy thickens slightly.

3 While the gravy is simmering, prepare the bubble and squeak. Heat the oil or butter in a skillet, add the cooked potatoes and fry over high heat, stirring regularly, until golden. Add the cooked cabbage and continue to fry over medium heat for 3 minutes, or until everything is hot. Do not allow the cabbage to burn. Season with salt and pepper, and serve with the sausages and gravy.

Chili-rubbed smoked pork belly with chipotle sauce

Rubbing fatty pork in aromatic spices before smoking gives it a characteristic flavor. Teamed with pickled vegetables and a fiery sauce, this is a satisfying dish to remember.

Serves 4

Ingredients

3 tablespoons (45 g) smoked paprika
1½ tablespoons (20 g) dark (hot) chili powder
2 tablespoons (30 g) brown sugar
2 teaspoons (10 g) salt
¼ teaspoon onion powder
¼ teaspoon garlic powder
¼ teaspoon ground cumin
½ teaspoon ground coriander
¼ teaspoon black pepper
¼ teaspoon white pepper
¼ teaspoon ground cinnamon
2 pounds (900 g) pork belly
apple or cherry wood for smoking

For the pickled vegetables
1 quart (1 l) water
2 cups (475 ml) sherry vinegar
3 tablespoons (45 g) salt
½ cup (125 g) sugar
2 tablespoons (30 g) black peppercorns
2 tablespoons (30 g) coriander seeds
2 cups (about 350 g) purple or white cauliflower florets
12 pearl onions, peeled
4 cherry bomb (Peppadew) peppers
4 heads baby fennel

For the chipotle sauce
2 chipotle chiles in adobo sauce
¼ cup (about 30 g) roasted garlic
juice of 1 lime
1 cup (240 ml) mayonnaise
salt and black pepper

Method

1 Mix together the herbs, spices, and other flavorings, then rub well into the pork belly. Cover with plastic food wrap and chill for 2–3 hours, or overnight to marinate.

2 When you are ready to cook the meat, smoke it over apple or cherry wood for 30 minutes. While it is smoking, preheat the oven to 300ºF (150ºC). Remove the meat from the smoker and roast in the oven for 1–1½ hours, or until tender.

Chef's Tips

Smoking meat adds flavor and tenderizes the flesh. Always use a smoker specially designed for the purpose, in a well-lit and well-ventilated area, and be sure to follow the manufacturer's instructions.

3 To make the pickled vegetables, put the water, vinegar, salt, sugar, peppercorns, and coriander seeds into a large pan and bring to a boil. Simmer for 3–4 minutes. Add the vegetables and continue to cook until the liquid reaches 185˚F (85˚C) on a candy thermometer. Remove from the heat and leave to cool.

4 To make the chipotle sauce, place all of the ingredients, except the seasoning, in a blender and purée until smooth. Season with salt and pepper, to taste.

5 Carve the meat into slices. Serve with the pickled vegetables and a drizzle of the chipotle sauce.

101

Roast pork shoulder with rhubarb & ginger sauce

Rhubarb and ginger are a great flavor combination and naturally cut through the richness of the meat.

Serves 4

Ingredients

2 onions, peeled and coarsely chopped
2 carrots, peeled and coarsely chopped
salt and black pepper
8–9 pound (3.5–4 kg) pork shoulder roast, on or
   off the bone—ask your butcher to score the fat
1¼ cups (300 ml) water

For the sauce
2¼ pounds (1 kg) rhubarb, chopped
1 inch (2½ cm) peeled, chopped ginger, or 2 pieces
   crystallized ginger, or 2 teaspoons (10 g) ground
   ginger
3 tablespoons (45 ml) water
½–1 cup (125–225 g) sugar

Method

1 Preheat the oven to 300°F (150°C). Place the onions and carrots in a Dutch oven or a casserole dish. Season the pork with salt and pepper, then place it over vegetables. Add the water. Cover and bake for 3½ hours, adding more water if necessary to prevent the meat from drying out.

2 Uncover the meat. Pour off and reserve any liquid. Increase oven temperature to 400°F (200°C). Roast the pork, uncovered, for 1½ hours, or until the skin is crisp and the meat is falling apart. Take care not to let the pork skin burn, and cover loosely with foil if necessary.

3 While the pork is cooking, make the sauce. Place the rhubarb, ginger, and water in a pan. Cover and simmer for 10 minutes, or until the rhubarb is soft. Add sugar to taste.

4 Remove the pork from the oven. Cover with foil and leave to rest in a warm place for 45 minutes.

5 Pour the reserved meat juices into a pan. Stir over medium heat, breaking up the vegetables to thicken the juices and make a gravy. Serve the gravy as it is, with the pork and rhubarb sauce, or push it through a sieve for a smoother texture.

Stuffed pork loin chops

A simple stuffing, flavored with thyme and lemon zest, gives these juicy chops a tasty twist. To keep the stuffing in place while the chops cook, you can seal them with a toothpick.

Serves 4

Ingredients

1½ tablespoons (25 g) chopped fresh thyme
2 tablespoons (20 g) grated lemon zest
2–3 tablespoons (30–45 ml) light olive or sunflower oil
4 butterflied top loin pork chops (pork loin steaks),
   4 ounces (115 g) each, trimmed of fat
¼ cup (25 g) freshly grated Parmesan cheese
¾–1 cup (100–120 g) fresh bread crumbs
salt and black pepper
1 egg, beaten

Butcher's Tips

You want an even amount of meat on each side of the chop, so ask your butcher to do the butterflying for you if you're not sure your knife skills are up to this delicate job!

Method

1 Combine ½ tablespoon (5 g) of the thyme, 1 tablespoon lemon zest, and 1½ tablespoons oil, and rub into the chops. Place in a nonmetallic dish and cover with plastic food wrap. Chill for 2 hours, or overnight, to marinate.

2 When you are ready to cook the chops, preheat the oven to 375°F (190°C). Mix together the remaining dry ingredients and season to taste. Stir in the beaten egg, a little at a time, until the mixture comes together—you may not need it all.

3 Open out the chops and spread with the stuffing on one side, leaving a ½-inch (1-cm) margin all round. Close the chops and press the edges together. Seal with a toothpick or tie loosely with kitchen string to keep them closed.

4 Place a skillet over high heat. Brush the chops with a little oil and cook in the hot pan for 1–2 minutes on each side, or until just golden. Transfer to a roasting pan and roast for 20 minutes, or until the meat is cooked through. Remove toothpick or string before serving.

Slow-cooked pork carnitas

This tough cut needs many hours in the oven to turn tender, but then you're ready for a Mexican-style feast where everyone can prepare their own tortilla at the table.

Serves 4

Ingredients

1 tablespoon salt
2–3 pounds (900 g–1.3 kg) pork shoulder (butt)
2 strips orange peel
3 cloves garlic, smashed and peeled
1 jalapeño pepper, seeded and cut in half
1 cinnamon stick
2 bay leaves
½ teaspoon cumin
3 teaspoons (15 g) pink peppercorns
2 quarts (2 l) water

To serve
corn tortillas, sliced avocado, chopped cilantro (coriander), fresh salsa, crumbled cheese, and lime wedges

Butcher's Tips

The butt comes from the front end, not the rear, of the pig. It's the shoulder, and the term "butt" comes from the barrels, known as butts, in which it was once shipped.

Method

1 Preheat the oven to 375°F (190°C). Rub salt all over the pork and let stand at room temperature for 15 minutes.

2 Place the pork in a Dutch oven or casserole dish. Add the remaining ingredients. You may need to add more water—it should come two-thirds of the way up the meat.

3 Cover the pot with a lid and cook in the oven for 3 hours, or until the meat is tender enough to shred with a spoon. If the meat is dry and tough, add a little more water and cook for another 30 minutes. When completely tender, transfer to a cutting board and shred meat.

4 Serve the meat wrapped in corn tortillas and topped with slices of avocado, chopped cilantro, salsa, and crumbled cheese. Add a squeeze of lime juice on top.

chapter 3

Lamb

Exceptionally hardy and able to thrive on marginal land, sheep have been farmed for millennia, not just for their succulent meat but also for the valuable wool and nutritious milk that they provide.

From Farm to Table

Traditional sheep husbandry encompasses three separate economies—meat, wool, and milk. Spring lamb meat remains at the heart of this rural industry, although older animals offer unbelievably tasty meat when cooked with care.

SYMBOL OF SPRING

The ultimate symbol of seasonal rebirth, lambs are usually born in springtime. They are nursed by their mothers, then weaned at eight weeks. Most move onto open pasture or continue on vitamin-enriched grain, plus hay. Intensively reared lamb is grown to maturity in specially prepared lots where they receive formulated feed. If the quality of the pasture is poor, grass-fed lamb are also fed grain in order to gain weight during the last 30 days before slaughter.

Farmers have developed methods to enable breeding ewes to give birth more often than their natural annual cycle. This allows the market to offer supplies of fresh lamb meat through the year, although the practise is frowned upon by some because it does not give ewes a natural break between births.

NICHE MEAT

Lamb consumption has fallen in the U.S. over recent decades, although its reputation as a prime meat with unique qualities and unrivalled culinary opportunities has never been questioned. Prices remain high, making lamb a considered choice for festive celebrations, rather than an everyday meat. However, since lamb and mutton are common in Middle Eastern cuisines, much of the consumer demand comes from these specific communities.

Lamb organ meats are a delicacy in the U.K. and across Europe. Scottish haggis is one of the most famous dishes. The organ meats are ground with oatmeal, suet, onion, and herbs, then stuffed into a lamb or ox intestine and boiled. The dish is honored with a bagpipe toast during the national festival of Burns Night and served with parsnips and potatoes.

Salt Marsh Lamb

A revered French and English delicacy, salt marsh lamb comes from sheep that graze on coastal marshes where the estuaries are washed by the tides. Depending on the locality, the sheep graze on a variety of salt-tolerant grasses and herbs, such as samphire, sea lavender, sorrel, and sparta grass. These different foods give the meat subtly different flavors.

Age and Flavor

Lamb is sold at different ages throughout the year. Young lambs are slaughtered at six weeks prior to weaning. These pale-fleshed animals are a prized delicacy and are often cooked on a spit. Spring lamb refers to those animals born in winter and slaughtered before five months. This meat is favored for its tenderness—and is the most expensive you can buy. Lamb slaughtered in the fall comes from older animals—it has a deep pink color and flavor and firmer fat.

OLDER AND TASTIER

Yearlings, or hogget, are sheep that are between one and two years, while mutton are usually aged two or older. Mutton is often derived from culled breeding rams and ewes, and is sold at a low return or bred specifically for the specialty market.

Cheaper to buy, yearlings and mutton offer a unique repertoire of recipes and are traditionally eaten during fall and winter when lambs are unavailable. Older sheep meat has a higher concentration of fatty acids, which alters the flavor and texture. With more connective tissue, the meat has a tighter texture and a gamey flavor. It suits slow-cooked and highly seasoned stews and pot-roasts. Mutton even has its own fan club, with Prince Charles as its patron. In England, the Mutton Appreciation Society recommends buying fresh mutton between October and March because the sheep have access to summer and autumn pastures, allowing them to gain extra weight. The animals are also finished on root crops and silage to prepare them for market. The darker the color, the older the animal. Baby lamb meat is pale pink, while Spring lamb is pinkish-red, and mutton is a deep shade of burgundy.

Across the world, most lamb is raised on pasture, but in the U.S., much is reared intensively. Grain-fed American lamb has a mild flavor, contains a high percentage of fat, and is butchered into larger-than-average cuts. British, Australian, and New Zealand lamb is mainly grass-fed and smaller in size.

According to some enthusiasts of pasture-only lamb, grain-finished lamb has excess quantities of gristly fat. Although there is more meat on the bone, the taste does not accept strong seasonings, and the meat texture is flaccid. The fat on grass-fed lamb is soft, the meat is dense, with a deeper flavor.

HANGING AND AGING

Lamb benefits from a short hanging period to age and mature the meat. The aging process is quicker for lamb than for beef. Traditional butchers hang lamb for about one week, although meat reared for the mass market gets less than three days. Meat reared specifically as mutton favors a two-week hanging.

Other Grill Cuts

All the lamb farmed today is derived from a few precious breeds. Many originate in England, and have been crossbred to create ideal meat lambs. Here are a few:

- **CHEVIOT** With its sweet meat, this small, Scottish, white-faced sheep thrives on poor pastures but still offers excellent meat.

- **HAMPSHIRE** This fast-growing breed is the source of flavorful, fleshy meat, and the lambs do not require grain finishing.

- **SHROPSHIRE** Originally favored for its dense wool, this dark-faced lamb offers leaner meat and a distinct taste when roasted, grilled, or pot-roasted.

- **SOUTHDOWN** As a docile breed, Southdowns are ideal for intensive farming. Leg cuts from the well-shaped carcass make a fine roast.

- **SUFFOLK** This large, black-headed sheep offers distinctly aromatic meat, both as lamb and mutton.

Leabharlanna Poibli Chathair Bhaile Átha Cliath
Dublin City Public Libraries

Grill & Broil

Lamb is a delicious, succulent choice for broiling and is also ideal for a barbecue over coals. But it pays to know your grilling cuts—go for tender meat cut from the leg or middle, rather than from the shoulder.

A fast, dry alternative to frying, the grill or broiler offers a blast of intense, red-hot heat that is ideal for tender lamb cuts. Grilling seals in the juices by forming a crucial crust on the surface of the meat. Lamb is such a great receptor of herb and spice flavors, especially garlic and rosemary, that most recipes recommend marinating or applying a tasty glaze to complement the meat and keep it moist. Oil the grill rack to prevent the meat from sticking and only turn the lamb once during cooking, as this will produce juicier results. Don't forget to rest the cooked meat before serving.

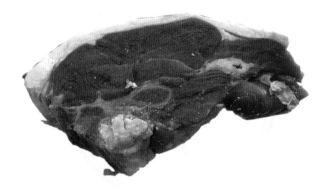

Sirloin Chop

A firm, meaty bone-in steak from the rump or leg. Ensure that the grill or broiler is thoroughly preheated before cooking for the all-important caramelization on the outside, and allow six–eight minutes cooking each side.

A.K.A. Chump chop

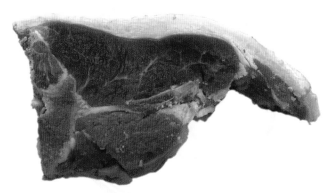

Loin Chop

A lamb-equivalent of a T-bone steak, this juicy cut has a generous layer of lubricating fat on one side that keeps it moist. Best cooked with the fat on, simply cut it off on the plate if preferred. Turn carefully during cooking to keep the meat in shape.

A.K.A. Sirloin chop

Rib Chop

A single chop from the rib, this meaty bone always does well on the grill and is fun to eat. They can be quite small, so allow two or three per person. Best cooked medium-rare in the center, as their leanness can make them dry. A French-trimmed rib chop has the meat removed from the slimmer part of the bone.

A.K.A. Cutlet

Other Grill Cuts

- **CENTER SLICE** Also sold—simply—as leg steaks, this cut is from the lean, harder-working part of the animal. Always marinate before grilling and take care not to overcook the meat. For best results, let the meat rest on a warmed plate before serving.

- **BUTTERFLY LEG** A barbecue favorite, this large, open, boneless leg steak needs at least 15 minutes on each side over a medium grill. Hold the meat firmly on the grill to ensure even cooking. Baste well with a herb or spice glaze. Ask your butcher to prepare the cut specially for you.

- **NECK FILETS** Fat-heavy, boneless meat, which is full of flavor and ideal for any form of fast grill cooking. Neck filet is sold ready-cubed for threading onto kabobs or pre-sliced for slender scallops (escalopes).

- **CUBED KEBAB MEAT** A lamb kebab is only as good as the meat you thread onto the skewer. Ask your butcher for lean lamb, cut into even-sized pieces. Without excess fat or connective gristle, the cooked meat will be easy to eat off the stick and absorb all the wonderful flavors of your marinade. Baste well while grilling.

Roast

Roast lamb is an occasion worth celebrating, and it is to this meat that we turn for festive feasts. Some cuts work best roasted in a super-slow oven, while others need the hot treatment.

L amb is a delicate meat with a soft texture. You can savor the satisfaction of carving a whole roast at the table into thick slices, full of infused taste and precious juices. Lamb roasts are seasoned before cooking by studding or rubbing with herbs or spices. Always reserve your meat ahead at busy holiday times, such as Easter. Loosely tie up larger cuts with kitchen string before roasting to maintain the shape of the meat and ensure even cooking.

Leg

The most welcoming roast, a single leg yields many portions of lean tender meat. Larger legs may be divided into two cuts. The sirloin (top half) or the shank (lower half)—ask your butcher for a cut to suit the number of guests. Legs are available bone-in (above) or boneless. Take care if you buy oven-ready boned legs as these may already be highly seasoned.

Rib Roast

The prime lamb roast, this pricey cut is cooked hot and fast, and served pink. A carefully butchered rack of lamb from both sides of the animal carcass can be joined together to form an impressive "crown roast." Separate rib chops also roast well.

A.K.A. Best end of neck, best end, rack of lamb

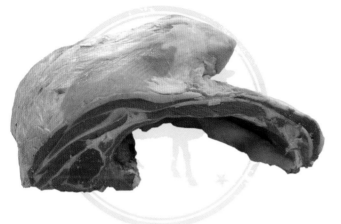

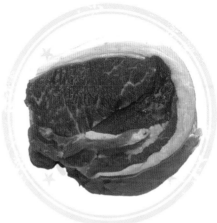

Shoulder

This economical cut enjoys very slow roasting (five hours plus) to let the fat melt into the marbled meat and impart succulence. The center bone makes this roast a job for a skilled carver, or you can tease the cooked meat away with a fork to serve. A good butcher can provide many cuts from one shoulder. It may be boned, stuffed, and rolled, separated into knuckle and filet roasts, or sliced into chops.

Noisette

Sliced from a boned-out loin, these rolled-and-tied cuts make neat little dinner bundles ideal for hot, quick roasting. The whole loin also makes a super-tender, but expensive, roast. It is commonly boned, stuffed, and rolled before cooking.

A.K.A. Loin chop

Other Roast Cuts

- **CENTER LEG** Taken from the top of the leg, this bone-in or boneless cut is also known as leg sirloin. It is fattier than the shank and cooks evenly because of its rounder shape.

- **SADDLE OF LAMB** This large, impressive cut comprises the entire loin, filet, and flank. A great challenge for new carvers, ask your butcher for assistance when ordering for a grand family celebration.

- **TOP ROUND LEG** A lean, medium-sized cut.

- **BUTTERFLY LEG** A boned leg, split open for stuffing and rolling.

- **BREAST** This economical, tender-fleshed, boneless joint suits a dry, herby stuffing.

- **SARATOGA ROAST** Also known as chuck-eye, this is a rolled, boned shoulder.

Braise & Stew

Cheaper cuts of lamb are a great choice for slow stews, homely pot-roasts, and braises. Select these cuts not just for traditional dishes, but for spicy tagines and curries, served with fluffy rice or flatbreads.

Slow cooking tenderizes firm or tough cuts and allows seasoned liquid to mingle with the meat. Braising and pot-roasting are best for seared whole cuts, while stewing is beneficial to smaller pieces of lamb. Braising usually requires that the level of liquid is no more than halfway up the side of the meat. This prevents the meat from drying out and provides enough liquid heat to break down the connective tissue and tougher flesh. As an extension of this technique, stewing fully immerses the meat in liquid, letting the moist heat permeate the flesh. All slow-cooked lamb is served well-done.

Neck Filet

Since lambs have pretty short necks, there isn't a lot of neck meat per animal. But when braised slowly, this marbled cut offers superbly sweet meat. Ask your butcher to trim off any sinew. Mutton neck also braises well, but cook for longer until tender.

Cubed Leg or Shoulder

Both the leg and shoulder are great boned and cubed for stewing. Due to slightly different cooking requirements—shoulder takes longer than leg—do not cook the two together. Trim off the excess fat, if preferred, and always sear the meat before stewing.

A.K.A. Diced lamb, stewing steak, stew meat

Breast

Extremely economical and very fatty, this is the lamb equivalent to pork belly. It is a long, thin cut that includes the wedge of fat from the top and lower part of the ribs. Ask your butcher for a single breast, the bigger the better. Roll and tie for braising, or leave flat and crisp up over the barbecue or under the broiler after braising.

A.K.A. Spareribs, Denver ribs

Other Braising Cuts

- **SCRAG END** The boniest part of the neck can be cut across into thick slices by your butcher. This cut has masses of flavor.

- **RIBLETS** Small rib cuts that take well to braising in Asian or Middle Eastern seasonings. Crisp up with a tasty glaze after cooking for a finger-licking meal.

- **BLADE, ARM, AND SIRLOIN CHOP** A single slice of shoulder or bone-in loin makes a convenient braising cut.

- **SHOULDER** A shoulder joint may be braised very slowly (five–seven hours) as a no-fuss alternative to roasting. The seasoned cooking liquor provides a superb, ready-made sauce.

Shank

This wonderfully meaty cut is ideal for pot-roasting and served with creamy mashed potatoes. After long, slow cooking, the meat is so tender that it simply falls off the bone. Ask your butcher to recommend the right number of shank cuts as they vary in size.

A.K.A. Shank end, fore shank, hind shank

Lamb kebabs

Kebabs are surprisingly quick and easy to make, especially if you ask your butcher to cube and trim the meat for you. Prepare well in advance and let the marinade transform the flavor.

Serves 4–6

Ingredients

1½ pounds (680 g) lamb sirloin, cut into
   1½-inch (4-cm) cubes
2 tablespoons (20 g) freshly chopped rosemary
   or 2 teaspoons (10 g) dried rosemary
2 teaspoons (10 g) ground black pepper
2 tablespoons (30 ml) soy sauce
¼ teaspoon ground cumin
grated zest and juice of 1 lemon
2 cloves garlic, peeled and smashed
vegetable oil, for greasing

To serve
warmed pita bread or couscous

Butcher's Tips

Marinating the lamb for an extended period of time—overnight if possible—imparts a whole lot more flavor, so you'll be rewarded if you plan ahead.

Method

1 Soak 8–12 wooden skewers in water for 30 minutes to prevent them burning in the oven.

2 Meanwhile, combine all the ingredients, except the oil, in a mixing bowl. Cover and chill for at least 15 minutes, but overnight if possible.

3 Preheat the oven to 375°F (190°C) and lightly grease a baking sheet. Bring the lamb to room temperature. Discard the garlic and thread 3–4 pieces of lamb onto each skewer, leaving room between each piece. For even cooking, skewer pieces of the same size together.

4 Spread out the skewers on the baking sheet. Bake, turning them once or twice during cooking, for 8–10 minutes for medium-rare. Serve in warmed pita bread or with couscous, accompanied by shredded lettuce, sliced tomatoes, and cucumbers.

Baked lamb meatballs

The special combination of herbs in these meatballs gives them a Greek flavor. Baking in the oven uses less oil than frying in a skillet—making them a healthier option, too.

Serves 4

Ingredients

olive oil, for greasing
½ onion, peeled
3 cloves garlic, peeled and smashed
2 slices day-old white bread, crusts removed
   and torn into pieces
1 large egg
¼ cup (60 ml) milk
handful of fresh parsley leaves
handful of fresh mint leaves
2 tablespoons (30 ml) olive oil
1½ teaspoons (7½ g) dried oregano
1½ teaspoons (7½ g) salt
1 teaspoon (5 g) freshly ground pepper
1½ pounds (680 g) ground lamb

To serve

warmed pita bread or flatbread, lettuce, tomato,
   cucumber, garlicky yogurt sauce, and hummus

Method

1 Preheat the oven to 375°F (190°C). Brush a baking sheet with olive oil.

2 Put all the ingredients, except the lamb, into a food processor and pulse until finely ground. Put the lamb into a mixing bowl, add the processed ingredients, and stir to combine. You may find it easier to do this job with your hands, wearing latex gloves.

3 Form the mixture into golf-ball-size balls and place them on the baking sheet, spaced ½ inch (1 cm) apart. Bake in the oven for 30–40 minutes until browned and cooked through. Serve with warmed bread, salad vegetables, garlicky yogurt sauce, and hummus.

Lamb broth with croquettes

Turn this warming broth into a dinner-party delight by serving it in elegant bowls and topping with a crisp goat's-cheese croquette.

Serves 6

Ingredients

For the broth
1 whole bone-in lamb shank
salt and black pepper
½ cup (100 g) pearl barley, washed and
　soaked overnight
8 carrots, peeled and finely chopped
3 onions, peeled and finely chopped
1 clove garlic, peeled and minced

For the croquettes
8 ounces (225 g) fresh goat's cheese
2 eggs
¼ cup (60 ml) milk
1 cup (125 g) all-purpose flour
½ cup (60 g) fresh bread crumbs
vegetable oil, for greasing or deep-frying

Method

1　Place the lamb in a large pot and fill with water. Season with salt and pepper. Bring to a boil, then simmer for 2–2½ hours, skimming off any residue regularly from the surface, until the meat falls off the bone. Add extra water to keep the meat covered. Lift out the meat and allow to cool.

2　Add the barley, carrots, onions, and garlic to the cooking liquid. Simmer for 50 minutes, skimming regularly. Top up with water as the barley swells, and cook until the vegetables and barley are tender. Shred the meat and add to the broth. Check the seasonings again.

3　For the croquettes, shape the cheese into 4 round portions. Break the eggs into a bowl and beat together with the milk to combine. Place the flour in a second bowl, and the bread crumbs into a third. One at a time, roll the cheese in the flour, then in the egg, and finally in the bread crumbs.

4　To cook the croquettes, either deep-fry in hot oil for 2 minutes, or bake for 5–7 minutes on a greased baking sheet at 375°F (190°C), until golden. Serve on top of the hot broth, with a slice of sourdough bread at the side.

Lancashire hotpot

This traditional lamb, onion, and potato stew from northern England is light enough for summer nights and takes just a few minutes to assemble. Serve with buttered carrots or broccoli.

Serves 4

Ingredients

2 pounds (900 g) boneless rib or shoulder, cut into
 1½-inch (4-cm) pieces
salt and black pepper
1 teaspoon (5 g) dried thyme
2 bay leaves
2 large onions, peeled and sliced
1½ pounds (680 g) potatoes, peeled and sliced
1½ cups (350 ml) lamb or vegetable stock
sunflower oil or melted butter, for glazing

Butcher's Tips

This dish would have originally been made with mutton meat, from sheep aged 1½ years and over. Chops from fully grown hill sheep would be stood upright around the edge of a tall ceramic pot, with the vegetables placed in the center. For a more rustic recipe, use bone-in chops as well as boneless meat.

Method

1 Preheat the oven to 325°F (170°C). Place the lamb in a Dutch oven or a large ovenproof casserole dish. Season with salt and pepper, sprinkle over the thyme and bay leaves. Add the onions, then arrange the potatoes on top in overlapping layers.

2 Slowly pour the stock into the dish. Brush the top of the potatoes with oil or butter. Cover and cook for 2½ hours, or until the meat is tender.

3 Increase the oven temperature to 400°F (200°C). Remove the lid and bake for another 30 minutes, or until the potatoes are golden.

Lamb two ways

This restaurant favorite is a snip to make at home. Neck of lamb is slowly braised with red wine, shallots, and mushrooms, then served with additional slices of rare roast lamb.

Serves 6–8

Ingredients

2¼ pounds (1 kg) cubed neck filet or shoulder meat
salt and black pepper
½ cup (60 g) all-purpose flour
4 tablespoons (60 ml) vegetable oil
2 cups (475 ml) pinot noir red wine
7 ounces (200 g) thick-cut smoked bacon, cut
   into strips (lardons)
20 small shallots, peeled
20 button mushrooms
1 sprig rosemary
6 sprigs thyme
2 tablespoons (30 g) red currant jelly
1¼ pounds (680 g) lamb leg or rump
1–2 tablespoons (15–30 g) gravy browning or gravy
   granules, optional

To serve
dauphinoise potatoes and green beans

Method

1 Season the cubed lamb meat, then toss in half of the flour. Heat 1 tablespoon of the oil in a skillet. When the pan is smoking hot, add the lamb in batches, and fry until golden all over. Set aside.

2 Use ½ cup (120 ml) of the wine to deglaze the pan and add the liquid to the meat. Heat 1 tablespoon of the oil in the skillet. Add bacon and fry until golden. Set aside in a bowl. Sauté the shallots and mushrooms in another tablespoon of oil until soft. Stir in ¼ cup (60 ml) of the wine, then add the cooked vegetables to the bacon.

3 Put the cubed lamb meat, remaining wine, and herbs into a large pot. Add enough water to cover the meat. Simmer very gently for 1½ hours, regularly skimming and reserving the fat. Add extra water if necessary to keep the meat moist.

4 Add the bacon and vegetables to the pot. Cook gently for 45–60 minutes, skimming regularly, until the meat is tender. Strain the meat and vegetables and set aside. Heat 1 tablespoon reserved lamb fat in a pan. Stir in remaining 1 tablespoon flour to make a roux. Whisk in the strained liquid until smooth, then add the red currant jelly. Heat, whisking, until thickened. Season, then add reserved meat and vegetables.

5 For the roast lamb, preheat the oven to 400°F (200°C). Season the lamb leg or rump, then seal in a hot skillet, in the remaining 1 tablespoon oil, for 5 minutes, or until browned. Transfer to a roasting pan. Roast for 12 minutes, then allow the meat to rest, covered, for 5 minutes.

6 Meanwhile, reheat the braised lamb. Use the gravy browning or granules to thicken and darken the cooking juices, if desired, following the packet instructions. To serve, carve the roast lamb and serve with the braised lamb, potatoes, and beans.

Butcher's Tips

Check the roast lamb towards the end of the roasting time. It should feel fairly firm when squeezed for medium-rare; if it is squishy it will be too rare, while hard or shrunken indicates well-done.

Lamb sausage with marble potato salad

Homemade sausages are nothing like store-bought—once you've tried them you will never go back! You can make the zingy salad with any kind of new potato, and it is also ideal at a barbecue.

Serves 4

Ingredients

1 pound (450 g) lamb shoulder, diced and chilled
5 ounces (140 g) cubed pork fatback, chilled
1½ tablespoons (20 g) salt
2 teaspoons (10 g) ground black pepper
2 teaspoons (10 g) minced garlic
2 teaspoons (10 g) chopped fresh rosemary
¼ cup (55 g) crushed ice or iced water
about 2 feet (60 cm) natural sausage casings
¼ stick (30 g) butter

For the marble potato salad

1 pound (450 g) red or white marble (small new) potatoes, washed
1 small red onion, peeled
2 piquillo chiles, seeded
2 teaspoons (10 g) whole grain mustard
3 tablespoons (45 ml) aged sherry vinegar
½ cup (120 ml) Spanish extra-virgin olive oil

Method

1 Grind the lamb and pork through the medium setting on a grinder or food processor. Chill for at least 30 minutes.

2 When cold, place the meat mixture, salt, pepper, garlic, rosemary, and ice or iced water back into the processor. Mix with the paddle (or in a bowl with your hands) for 1 minute, or until well incorporated. The ice or iced water makes it easier to stuff the sausages. Chill again.

3 Soak the sausage casings in warm water for 30 minutes, then fill with the meat mixture. Chill again for up to 1 day until ready to cook.

4 Meanwhile, make the salad. Steam the potatoes over boiling water until tender. Drain and cut in half. Slice the red onion and chiles into thin strips. Place in a bowl with the potatoes.

5 In a bowl, whisk together the mustard, vinegar, and olive oil. Pour over the vegetables, and stir until evenly coated.

6 To cook the sausages, melt the butter in a sauté pan. Add the sausages and fry gently until cooked through. Do not allow the pan to get too hot or the sausages will burst. Serve with the salad.

Chef's Tips

If you want to check the flavors before putting the sausage into the casings, fry a little in a skillet, taste it, then adjust seasoning until you are satisfied.

Agnello scottadito

Scottadito means "burned fingers" in Italian; this lamb smells and tastes so good that you'll be tempted to eat it straight from the pan! The meat is served dry and well-done.

Serves 4

Ingredients

4 lamb chops
4 slices lamb leg or rump steak
4 slices rib roast or best end of lamb
½ cup (90 g) dry (bitter) black olives
¼ cup (60 ml) olive oil
¾ cup (175 ml) red wine
4 cloves garlic
2 sprigs rosemary
1 red chile, chopped

To serve
sautéed or roasted potatoes

Method

1 Put all the lamb pieces into a large ceramic baking dish. Add the rest of the ingredients, mixing together gently. Cover the dish with plastic food wrap. Refrigerate for 4 hours, or overnight, to let the lamb marinate.

2 About 1 hour before you are ready to eat, preheat the oven to 425°F (220°C). Roast the lamb until the juices evaporate, but do not allow to dry out.

3 Serve the lamb with sautéed potatoes or roasted potatoes cooked in the oven with the lamb.

Chef's Tips

Bitter black olives are essential to the taste of this dish. To make your own, marinate black olives in olive oil overnight, then roast in the oven at 400°F (200°C) for 1 hour, or until they are dry and wrinkled.

Lamb tagine

Conjure up the flavors of Morocco with this traditional casserole. You can buy preserved lemons and spicy harissa paste in specialty stores.

Serves 4

Ingredients

2 tablespoons (30 ml) olive oil
2¼ pounds (1 kg) neck of lamb, or other good stewing meat, cut into 2-inch (5-cm) chunks
3 Spanish onions, peeled and sliced
3 cloves garlic, peeled and chopped
3 carrots, peeled and chopped into large dice
2 sticks celery, chopped into large dice
1 tablespoon (10 g) ground cumin
1 tablespoon (10 g) ground ginger
3 cinnamon sticks
1 tablespoon (15 g) harissa paste
2½ cups (600 ml) lamb stock or brown chicken stock
sea salt
2 preserved lemons, finely sliced
10 large green olives

To serve
cilantro (coriander) leaves and saffron couscous

Method

1 Heat 1 tablespoon of the oil in a skillet over medium heat. Add the lamb in batches and sear all over, taking care not to crowd the pan.

2 In a heavy casserole dish, sweat the onions in the remaining 1 tablespoon oil until nicely browned. Add the garlic, carrots, and celery, and cook for a few minutes. Add the spices and harissa paste, stirring all the time. Stir in the lamb stock. The sauce should be the consistency of heavy cream; if it is too thick add a little water. Season to taste with sea salt.

3 Add the lamb to the casserole, cover, and simmer very gently for about 1 hour.

4 Add the preserved lemons and olives, and cook for another 30 minutes, or until the lamb is tender. Serve garnished with chopped cilantro (coriander) leaves, accompanied by saffron-infused couscous.

Slow-roasted lamb shoulder

If you haven't tried a shoulder joint on the bone, then make sure you do! Cooked long and slow, the meat falls off the bone, and becomes so soft you can carve it with a spoon.

Serves 4

Ingredients

1 tablespoon (10 g) chopped fresh rosemary
1 tablespoon (10 g) chopped fresh thyme
3 cloves garlic, peeled and crushed
salt and black pepper
¼ cup (60 ml) extra-virgin olive oil
4 pounds (2 kg) bone-in shoulder of lamb

To serve
roasted root vegetables

Butcher's Tips

Since lamb shoulder has a lot of fat, this 4-pound cut can cook for a very long time without drying out. It smells and tastes divine!

Method

1 Preheat oven to 250°F (120°C). Blend together the rosemary, thyme, garlic, salt, pepper, and olive oil to form a smooth paste.

2 Using your fingers, smear the paste over the lamb shoulder. Cover with foil, and roast for 4–5 hours, or until the lamb is tender. Remove the foil for the final 45–60 minutes of the cooking time. Serve with roasted root vegetables cooked in the oven with the lamb.

Trimming lamb

When you're hosting a dinner party or celebration meal, you want food that looks as good as it tastes. Serve a beautifully trimmed cut of lamb as the centerpiece and it's bound to impress. Remember that your butcher can do the trimming for you and have the meat oven-ready.

The upper rib cut (best end) is an ideal cut for a trimmed presentation. A single rack, with the rib ends trimmed of meat and fat, is ideal for a small roast. But combine two racks into a guard of honor or a crown roast, and you have a stylish way of serving a delicious cut.

PREPARING A RACK

When buying a best end of neck for a single rack, ask the butcher to saw through the chine bone (backbone) but to leave it in place. At home, place the lamb fat-side down. Using the point of a sharp knife, remove the yellow connective tissue from under the rib ends. Turn over the rack and make a slit at the shoulder end so that you can remove any remaining shoulder blade. Now score a line across the rack 2–3 inches (5–7.5 cm) from the tips of the ribs, cutting through the fat and meat, down to the bones. Using a knife to help you, peel back the meat and fat to reveal the bones. Cut out the meat and fat from between each rib and freeze the trimmings for stock. Scrape off any meat and fat from the ends of the bones.

GUARD OF HONOR

A guard of honor is created by standing two racks facing each other, with the fat on the outside and the trimmed bones interlocking (see right). Ask your butcher for racks from the same animal so that they match and form a symmetrical roast. Also ask your butcher to remove the chine bones, and trim the racks. Stand the racks upright so that the fat is outermost and press together to interlock. To make sure the racks don't part company during cooking, tie kitchen string along the joint.

CROWN ROAST

A crown roast is formed from two racks of lamb, each curved into a semi-circle and then joined together form a ring. First prepare and trim the racks as described, then carefully bend each one into a semi-circle so that the meat is on the inner curve. Use kitchen string to tie the ribs together. Either tie the string around the bones or pierce a hole in the meat and thread the string through. Because the trimmed ribs on a guard of honor and a crown roast can blacken during cooking, it is traditional to serve these roasts topped with miniature paper chef's hats on the top of each bone. Alternatively, you can protect the tips of the ribs by wrapping each one in foil when roasting; remove 30 minutes before the end of cooking.

Frenching

Removing the meat, fat, and other tissue from the ends of rib bones is known as Frenching or French trimming. Rack of lamb, guard of honor, and crown roast are all trimmed before roasting. You can also trim individual lamb cutlets (chops). Simply cut off the excess fat and meat 2 inches (5 cm) from the end of the bone. Scrape away any membrane or connective tissue to leave the bone completely clean. This technique is ideal if you're preparing cutlets for a barbecue or buffet meal, when diners may want to pick up the cutlets with their fingers.

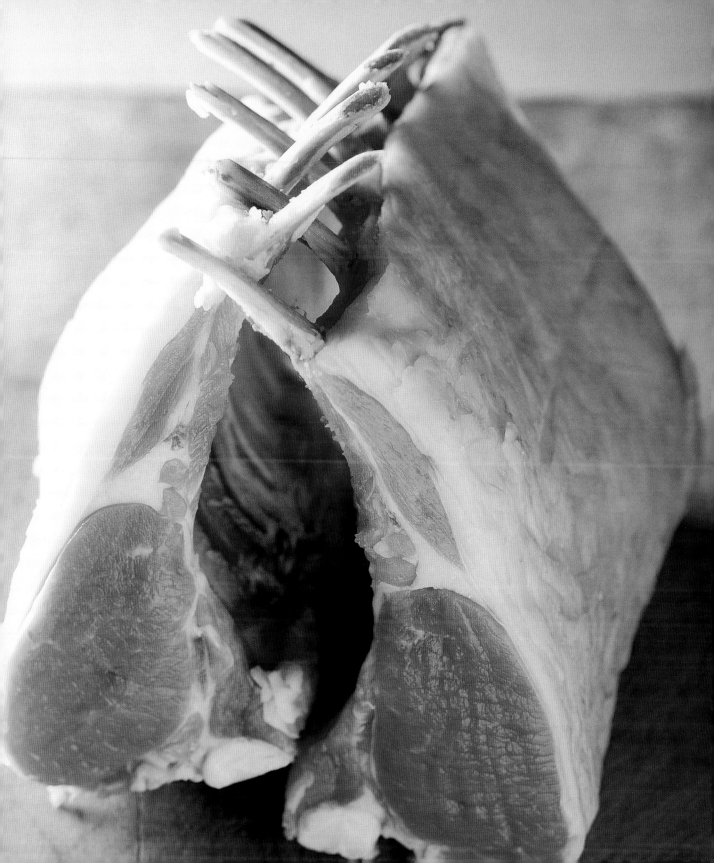

Roast rack of lamb with summer vegetable ragout

This classic French dish is a show-stopper at any dinner party. If you don't have time to trim the meat yourself (see page 128), simply ask your butcher to do it for you.

Serves 4

Ingredients

1 tablespoon (10 g) chopped fresh thyme
1 teaspoon (5 g) chopped fresh rosemary
¼ cup (60 ml) extra-virgin olive oil
2 French-trimmed racks of lamb, each weighing 2 pounds (1 kg)
salt and black pepper

For the summer vegetable ragout

3 tablespoons (45 g) butter
4 red onions, such as Tropea, peeled and quartered
1 tablespoon (15 g) minced garlic
¾ cup (120 g) cooked cranberry (borlotti) beans, or other beans, such as fava (broad) beans
¾ cup (120 g) fresh cooked crowder peas, or other peas, such as English peas
½ cup (55 g) Brussels sprouts, split into leaves
salt and black pepper
⅓ cup (80 ml) lamb stock or brown chicken stock
2 tablespoons (20 g) chopped fresh chives
1 tablespoon (10 g) chopped fresh parsley

Method

1 Mix together the herbs and oil. Coat the lamb with the mixture, then marinate in the refrigerator for 2–3 hours, or overnight.

2 Preheat oven to 400°F (200°C). Season the lamb with salt and pepper, then sear in a hot sauté pan until browned all over. Transfer to a baking pan. Roast in the oven for 8–10 minutes, or until medium-rare. Allow the meat to rest, covered, in a warm place for 6–8 minutes before serving.

3 While the lamb is resting, make the vegetable ragout. Melt 2 tablespoons of the butter in a sauté pan. Add the onions and cook gently until golden. Add the garlic, cooked beans, peas, and Brussels sprout leaves. Season with salt and black pepper. Sauté for 4–5 minutes, or until the onions are tender. Stir in the stock and bring to a boil. Lower the heat and stir in the final tablespoon of butter, followed by the chives and parsley. Serve immediately with the rested lamb.

Anchovy lamb with cauliflower "couscous"

Anchovies and lamb are a classic Italian combination—the saltiness of the fish brings out the sweetness of the meat. Serve with hazelnut-cauliflower "couscous."

Serves 4

Ingredients

1 onion, peeled and chopped
1 bulb fennel, cored and chopped
2¼ pounds (1 kg) ripe plum tomatoes, chopped
1 head garlic, broken into cloves
small bunch fresh thyme
3 sprigs rosemary
3 pounds 5 ounces (1½ kg) bone-in leg of lamb
10 cloves garlic, peeled and sliced
salt and black pepper
¾ cup (90 g) capers
3½ ounces (100 g) anchovies, drained
2 tablespoons (20 g) ground cumin
2 tablespoons (20 g) ground coriander
¼ cup (60 ml) olive oil

For the "couscous"

1 cauliflower, cored and broken into florets
zest and juice of 1 lemon
bunch of cilantro (coriander), chopped
¾ cup (90 g) peeled, toasted hazelnuts, finely chopped
1–2 cups (250–500 ml) olive oil

Chef's Tips

To save time, skip the gravy-making at step 6. Simply serve the rested lamb with the vegetables and juices straight from the roasting pan.

Method

1 Preheat the oven to 160°C (315°F). Place the onion, fennel, tomatoes, garlic cloves, and herbs in a roasting pan. Place the lamb on top. Use a sharp knife to make small incisions in the flesh, and insert the sliced garlic. Season to taste with black pepper.

2 In a food processor, blend the capers, anchovies, and spices into a coarse paste, adding just enough oil to bind the ingredients together. Smother the top of the lamb with the mixture.

3 Roast the lamb for 1–1½ hours, or until a meat thermometer registers 130°F–140°F (55°C–60°C).

4 While the lamb is roasting, make the cauliflower "couscous." Grate the cauliflower on a medium grater into a bowl. Add the lemon zest and juice, cilantro, and hazelnuts. Add enough oil to bind together, then season with salt and pepper.

5 When the lamb is cooked, cover with foil and allow to rest for 1 hour in a warm place.

6 For the gravy, put the vegetables and juices from the roasting pan into a colander and press with the back of a spoon to extract all the liquid. Discard vegetables. Pass the liquid though a fine sieve, then boil in a pan until reduced to the consistency of heavy cream. Carve the lamb and serve with the cauliflower "couscous" and gravy.

133

Roast shoulder of lamb with mustard & thyme

This succulent shoulder of lamb is flavored with aromatic mustard and thyme, and accompanied by delicate bundles of green beans wrapped in paper-thin pancetta.

Serves 6

Ingredients

1 bone-in shoulder of lamb, about 3 pounds
   5 ounces (1½ kg)
salt and black pepper
3½ ounces (100 g) whole grain mustard
1 tablespoon (10 g) fresh thyme leaves
6–12 cloves garlic 1 cup (240 ml) lamb or vegetable
   stock

For the beans

12 ounces (340 g) green beans, topped and tailed
12 thin slices pancetta
1 tablespoon (15 ml) olive oil, for brushing

Method

1 Preheat oven to 400°F (200°C). Place lamb in a deep roasting pan and season with salt and pepper. Mix together the mustard and thyme, then spread over the lamb. Add the unpeeled garlic to the pan.

2 Bring the stock to a boil, then pour into the pan. Cover loosely with foil so that the foil does not touch the meat. Roast for 1 hour. Baste the lamb with the pan juices. Roast, uncovered, for 30–45 minutes, or until the meat and tender. Rest the lamb, covered, for 15 minutes in a warm place.

3 While the lamb is roasting, blanch the beans for 3–4 minutes in boiling water. Drain and cool under cold water. Divide beans into portions, then wrap pancetta around each bundle. Place in a baking pan and drizzle with oil. Bake for 10–15 minutes, or until the pancetta is crisp.

4 Carve the lamb and serve with the garlic from the roasting pan and the wrapped beans. Skim off the excess fat from the roasting juices and serve the remaining juices on the side.

134

chapter 4
Game

Low in cholesterol and high in protein, what more can you ask from a meat? Game offers these best seller qualities, plus a smack of wild outdoor flavor. Your butcher can supply the freshest cuts and will advise you how best to prepare them.

From Field to Table

Hunting game is an ancient tradition that expresses the human need to capture food for survival and indulge in sporting pleasure. Hunting brings a superb range of wild meats to the dinner table.

American Game

The U.S. is a hunter's paradise, as the varied geography offers a wide variety of animals. Each state has its own strict hunting laws, and all game is killed in accordance with regional regulations. For reasons of both legality and food safety, take care to buy game from a butcher who can inform you of the full provenance of the animal and its post-hunt handling. Game raised on farms under appropriate regulations can be sold through a butcher or game dealer. Wild game hunted legally under federal or state authorities cannot be sold, but can be harvested for personal consumption.

DEER

Deer types vary (see page 141), but the most common type is the whitetail. A century ago, these deer numbered about half a million, but are now estimated at 20 million strong. The population explosion was caused by the demise of natural predators and hunting restrictions. Hunting controls the deer population and the environmental damage that they cause.

MOOSE AND ELK

The moose is the largest member of the deer family and lives in northern states. Males weigh about 1,500 pounds (680 kgs) and females around 900 pounds (408 kg). Due to their natural northern habitat, they are popular in Nordic cuisines—ground for burgers or meatloaf, marinated and grilled, or salted and dried. The fats and outer membranes ("silver skin") are removed when the animal is butchered, then individual cuts are prepared for roasts, ribs, and steaks.

Crossbred Boars

Wild boar in the United States are a genetic mix of wild Eurasian pigs and domestic American pigs. They are reared free-range and also hunted in the wild. The meat is best marinated before cooking— wine or pineapple juice tenderize the flesh perfectly.

JAVELINA (PECCARY OR MUSK HOG)

Found in Texas, Arizona, and New Mexico, javelina is a fast-breeding, pig-like mammal that lives in herds. The smaller females are best for cooking. As with all furred game, the hunter should shoot only in the head or neck to prevent damage to the body meat. The animal has a strong-smelling musk gland and thick hairs that must be removed carefully when butchered immediately after harvesting. The cuts are marinated before cooking or ground and mixed with pork.

BOAR (FERAL HOG AND PIG)

Hunting wild boar and feral hog is a popular sport. Many states have unregulated herds of nuisance animals, with no hunting or trapping restrictions. The meat is dark, lean, with a tight, compact grain. Check that the soft fat has been trimmed when jointed, and always marinate (wet or dry) the meat before cooking. Butchers do not sell feral pig because the health of the animal, and its handling after death, cannot be guaranteed. Source the meat from a local game hunter.

Choosing and Buying

If game is not shot cleanly in the head, the appearance, flavor, and texture of the entire carcass may be spoiled. Closely check the condition of the meat before you buy.

Remnants of lead-based ammunition are another possible hazard. Pregnant women and children should eat only a limited quantity of furred game (and game birds) because of the potential health risk from lead shot in the meat. Marinating or cooking the meat in wine-, vinegar-, or tomato-based sauces, which are acidic, allows the lead to dissolve and be absorbed more easily. Most lead shot is removed professionally during butchering and preparation, but tiny, invisible fragments can become embedded in the flesh and remain after cooking. NOTE: This warning is only relevant to truly wild game because mainstream game, such as venison sold by supermarkets, is domestically farmed and slaughtered in a conventional manner.

Wild boar are hardy, adaptable, and strong-willed creatures that cannot be tamed or domesticated. Expert marksmen hunt the animals for meat and hide.

More Exotic Meats

- **BUFFALO (INCLUDING BISON AND MUSK-OX)** Wild and farmed buffalo is an ideal alternative to beef. It is very lean, with a sweet delicate flavor. Domestic herds are raised for their creamy milk and tender meat.

- **CARIBOU (REINDEER)** With a lighter taste than other game, this lean meat is best cooked medium-rare. You can also marinate and braise as a stew or grind for a meatloaf.

- **GOAT** Goat is a delicious meat that is eaten across the Caribbean, Africa, the Middle East, and Asia. It suits slow braising to make it tender and takes well to strong spices. Goat meat is sourced from the dairy industry and the mustering of feral animals.

- **ANTELOPE** African black buck and nilgai antelope roam on huge Texan game preserves where they are raised for meat and hide. The lean meat taste like deer. Pronghorn is a similar meat and lives wild in Wyoming.

- **KANGAROO** Imported from Australia, kangaroo meat is similar to venison in flavor and very lean. For best results, grill steaks medium-rare or rare on a very hot barbecue.

- **ALLIGATOR** This strictly regulated game offers a range of tasty cuts, tasting a little like veal or chicken. The white tail meat, jaw, and backstrap cuts are extremely tender. The body and leg cuts are darker and are best ground or cubed for slow cooking.

Venison

No wonder venison is such a hit with meat-eaters—its wild country flavor and fine texture is both satisfying and stimulating. This deep-hued, free-range meat also offers less fat and cholesterol than beef.

Farmed and Wild

Deer are majestic animals that have long been prized for both sport and the unique meat that they supply. Venison meat—also known as roebuck—is classified according to its origin. It may be sold as either wild hunted deer, park deer that is reared on open land, or as farmed deer. Farmed deer is mostly free-range, but can also be raised intensively.

The best venison comes from truly wild animals, killed instantly with one shot during the natural hunting season. This age-old method does not give the animal any time to experience pre-slaughter stress, which induces adrenaline and risks damaging the texture and taste of the flesh.

MATURITY
The age of the deer determines the name given to it by its hunter or game dealer, and subsequently affects how it should be cooked. Up until the age of 18 months of age, the animal is called a "fawn." Between 18 months and two years, it is known as a "pricket" or "yearling" and, after that, a "brocket." Wise and experienced hunters know the age of a male deer by the number of branches on its antlers and, in the female, by the burrs on its head. This vital information should be passed to the butcher or game dealer.

Domestic farmed venison is reared year-round, usually from red deer, fallow deer, and elk. It is marketed as a healthy alternative to beef and sold mainly to restaurants and specialty suppliers once the meat has been inspected by the United States Department of Agriculture, or an equivalent authority. Farmed venison is taken from younger animals, resulting in a milder flavor and slightly fattier texture. Since venison is so lean, extra fat may be added when cooking to prevent the meat from drying out and to impart succulence.

HANGING, AGING, AND STORING
Butchers and abattoirs hang venison in a fully furred condition for between three days and three weeks. This process is crucial to relaxing the tough muscles. Deer are such lean animals that they dry out when skinned, which is why they are hung intact. The butcher takes control of the sanitation, temperature, humidity, and air circulation in the cold room. The entrails are removed as soon as possible after the animal is killed. The tenderest meat comes from smaller species and younger animals. A young roe deer may only need a few days aging, while an older fallow or red deer could need up to three weeks to mature before sale.

Fresh venison has a deep purple-red color, with closely packed meat fibers. There is little fat—virtually none on wild venison—but what there is firm and white. Never buy venison with yellow or flaccid-looking fat. The portioned meat may be refrigerated for two days if well wrapped and sealed. It can be frozen for one month, vacuum-sealed in freezer plastic. Venison also dries into the best-tasting jerky strips.

FLAVORS AND TASTY PARTNERS
Venison takes kindly to all traditional meat accompaniments and the fruity sauces that go well with other game meats, such as quince, apricot, and cranberry. Venison also loves being married with fresh horseradish, black pepper, or paprika, plus tasty strips of salty bacon or pancetta that moisten the lean meat as it cooks.

Deer Breeds

Wild deer are available at different times of the year depending on the natural breeding cycle of the animal and regional hunting seasons. State regulations vary, as do laws on firearms, age of the hunter, and the ownership of the hunting grounds. Check with your butcher or game dealer for the precise hunting seasons in your region.

▶ **Whitetail** A woodland species, widespread across the U.S., and the most commonly hunted deer.
▶ **Mule** This breed is indigenous to western states and has strict hunting restrictions due to falling numbers.
▶ **Red** Native to the U.K., this deer lives on wild moorland and feeds on herbs and heathers—a diet said to make it the best-tasting deer.
▶ **Fallow** Popularized by game hunters in the eighteenth century, the fallow deer interbreeds easily with Red deer, resulting in the dilution of its natural gene pool.

▶ **Roe** Reintroduced across Europe after being hunted to virtual extinction during the seventeenth century. Roe deer inhabit deciduous woods, surrounded by pasture.
▶ **Sika** An Asian species introduced to Europe and the Americas, Sika breed easily with other varieties. Found in Alaska, Wisconsin, Virginia, Kansas, and British Colombia.
▶ **Muntjac** This small Asian species is available all year in the U.K., having originally escaped from Woburn Safari Park back in 1925. It breeds easily in the wild.

What's in a Name?

Venison used to refer to all furred game. Now the term usually refers to deer, but can also include other horned game: caribou, elk, moose, and reindeer. Apart from deer, elk is the most common farmed game. Elk breeds well and is easily herded.

Early morning light settles on a whitetail deer in Wisconsin. The deer's internal organs are removed immediately after death, known as "field dressing." This essential process prevents the meat from spoiling and prepares the animal for hanging.

Venison Cuts

Although hunted and killed as game, venison is sold more like a domestic farm animal. Ask your butcher to show you the choicest venison cuts from the cold room, and use them to prepare the best-ever meat dinner.

A venison carcass is roughly the same size as a lamb. Simply select a single roast, or cubed meat for stewing. Fearing dryness, traditional venison recipes call for the addition of pork fat to compensate for the meat's leanness. However, if the meat is cooked simply in a moderately hot oven it can be tender and juicy and offer the pure taste of its wild, open habitat. It is not essential to marinate venison either: This process can draw out too many of its precious juices rather than retain succulence. Just remember to rest the meat sufficiently after cooking to let the fibers relax.

Cubed Stew Meat

Venison for stewing or braising comes from the neck, shoulder, shin, or shank. It is best cooked very long and super slow. The meat can also be ground for burgers and sausages, mixed with a little fatty pork to add succulence.

A.K.A. Stewing steak, diced venison

Filet Medallion

Let these tender steaks come to room temperature before cooking. Salt the meat and pat dry. Fry over medium heat for a minute or two, turning just once. Rest the meat for at least 10 minutes before serving.

A.K.A. Loin, tenderloin, backstrap, saddle steak, filet steak

Making the Most of Venison

Butchers and chefs are firm believers in utilizing the whole carcass—called "carcass balance." If cooks only choose to eat the prime cuts, the wonderful potential of marginal cuts is lost and precious food may be wasted. For example, the strong muscles of the deer can be cubed and used to make a wonderful pot-roast. The neck can be braised in an entire bottle of red wine, with pancetta and wild mushrooms to create a breathtaking main course.

Fresh venison organ meats are also a delight, especially the liver and heart. Sear over high heat and serve with fried eggs for a stunning hunter's breakfast. Venison liver is also ideal for pâté.

Saddle

The saddle is usually roasted on the bone and served pink in the center. The meat is rested for at least 20 minutes after roasting, as this helps finish the out-of-oven cooking process and retain succulence.

A.K.A. Loin

Rump Steak

Cook these slim steaks rare or medium-rare. If well-done, the meat will take on a livery taste, a rubbery texture, and a dry, stringy texture. After resting, the juices will look very red—this is not blood but the changing protein juices draining out of the flesh.

A.K.A. Topside, silverside, haunch

Rolled Haunch

Ideal for roasting or braising, this rolled, pork-barded boneless roast is cut from a whole round haunch and carves easily at the table. For a larger crowd, select a whole leg on the bone. Roast medium-rare, leaving it a little pink to keep it tender.

A.K.A. Leg roast

143

Hare & Rabbit

Hare and rabbit are wonderful sources of wild meat. They are affordable, healthy, and back on the menu at the best restaurants in town.

Farmed rabbit is widely available, but with so many animals in the wild there is really no need to buy farmed meat. Farmed rabbit is composed entirely of white meat, and milder than true wild rabbit, which has a stronger taste and a dark hue to the flesh. Wild rabbit offers a unique herby tang that is testament to its diet of natural, foraged vegetation.

THE BUTCHER'S SKILL

Killing and preparing rabbit or hare requires a degree of skill, so they remain the realm of the hunter and trained butcher, rather than the factory farm and supermarket.

Prior to butchering, a whole, skinned rabbit or hare reveals its distinctive shape and anatomy. If shot, it may also show a dark patch of blood indicating the place where it was killed. A good amount of bacon, pancetta, or other fat is usually added to make it tender and moist on the palate. Yogurt, cream, and even coconut milk can serve this purpose as well. The bones are easily removed before cooking—many of the most famous recipes use boneless cuts. Use a very sharp knife, slowly and carefully, when handling these small mammals.

COOKING TECHNIQUES

Fast, hot roasting is risky with hare and rabbit—it can work with the saddle of young, bacon-barded rabbits, but you need to be sure of the animal's age, or it may come out tough. A moist stuffing is an ideal addition to retain succulence.

The back legs of farmed rabbits can be roasted because they are a little fattier and do not dry out so easily. Braising is the safest option for rabbit, but check the tenderness of the meat as it cooks. Young rabbits can be ready in an hour, but an old buck on a low heat may take two hours.

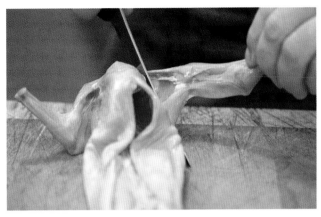

◀ 2. BACK LEGS

Now cut off the two larger rear legs. This can be tricky, so make sure your knife is super sharp. Hold the leg taut with one hand, then cut into the meat just below where the saddle ends. Keep the knife close to the hip bone. When the meat is cut away, the leg bone will just twist off. The legs are usually cooked as one piece.

▲ 1. REMOVAL OF FRONT LEGS

At the front of the carcass, you will see two small front legs, sometimes called the arms or shoulders. Using a sharp knife, cut off the front legs as close as possible to the body. You do this by working the knife along the joint between the leg and rib cage, until the joint comes clean away.

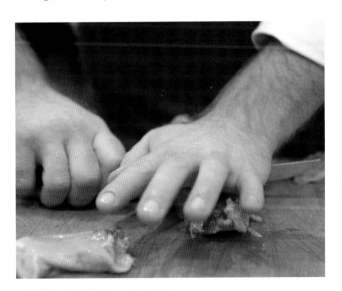

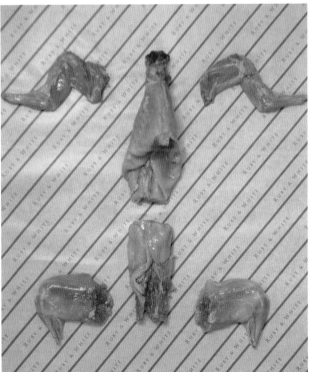

▲ 3. DIVIDE THE BACK

Cut through the backstrap meat to separate the rib section from the saddle. With the meat cut, press down firmly on the joint with your hand to snap and break the backbone. Use a clean cut through the bone to separate the two parts.

▲ 4. TRIM

With six, pot-ready rabbit portions, you can trim off the neck, abdominal flaps, and leg bones—keep these to enhance your stock or gravy. The saddle that lies between the two back legs can be roasted whole or split down the center.

Rabbit rillettes niçoise

Bring the South of France to your dinner table with this fresh homemade pâté. Serve with a crusty baguette, mustard, and a pile of cornichon pickles.

Serves 4

Ingredients

4 large, freshly braised rabbit legs, about 2 pounds (900 g) in total (see Butcher's Tips, below)
1 cup (about 225 g) rendered duck fat
1 tablespoon (15 g) minced fresh garlic
1 tablespoon (10 g) chopped fresh rosemary
1 teaspoon (5 g) lemon zest
¼ cup (50 g) pitted niçoise olives, finely diced
salt and black pepper

To serve

toasted baguette, whole grain mustard,
 and cornichons

Butcher's Tips

With their robust flavor and texture, the best cut for this coarse pâté is rabbit hind legs. Ask your butcher for evenly sized legs. First brown the meat in a little hot oil and butter, then braise in a low oven with about 3 cups (700 ml) stock for 45–60 minutes, or until tender.

Method

1 Pick the braised meat off the bones—it will be tender and come away easily. Put the meat in a food processor and process briefly until chopped.

2 Melt the duck fat in a heavy pan and bring to a very low simmer. Remove from the heat and add the garlic. Process the rabbit meat again until coarsely ground, gradually adding the fat to create a rough paste. Do not over-process.

3 Put the rabbit mixture into a bowl, and gently fold in the rosemary, lemon, and olives. Season to taste with salt and black pepper. Serve immediately with toasted baguette, mustard, and cornichons, or cover and chill, and use within 2 days.

Rabbit pie

A real old-fashioned dish, this tasty meat pie makes the most of rabbit. The rustic filling is enhanced with the flavors of white wine, mustard, and tarragon.

Serves 4

Ingredients

For the pastry
2½ cups (350 g) self-rising flour, plus extra for dusting
¾ cup (175 g) suet or vegetable shortening
salt and black pepper
2–3 tablespoons (30–45 ml) iced water
1 egg, mixed with 1 tablespoon (15 ml) milk

For the filling
2 rabbits, cut into 6 parts each
¼ cup (60 ml) olive oil
4 large shallots, peeled and sliced
2 cloves garlic, peeled and finely chopped
2 slices bacon, chopped
1 tablespoon (15 ml) Dijon mustard
1 cup (240 ml) white wine
1 cup (240 ml) chicken stock
⅓ cup (80 ml) heavy cream
1 tablespoon (15 g) chopped fresh tarragon
salt and black pepper

Chef's Tips

Before baking, make a couple of incisions in the pastry top to let the steam escape and stay crisp.

Method

1 For the pastry, put the flour, fat, and seasoning in a food processor. Pulse for 30 seconds, then add water to form a dough. Roll out to ¼ inch (6 mm) thick and cut around pie dish to make a lid. Chill.

2 For the filling, season the rabbit with salt and pepper. Heat half the oil in a heavy frying pan until smoking hot. Add the rabbit and sear on all sides. Transfer to an ovenproof casserole dish.

3 Preheat oven to 350°F (180°C). Sweat the shallots and garlic in the remaining oil until soft. Add the bacon and cook for 2 minutes. Add mustard, wine, and stock. Simmer until reduced by a third. Pour mixture over rabbit. Bake for 1 hour, or until the meat falls away from the bone. Remove the rabbit, reserving the cooking liquid. Allow the meat to cool, then strip it away from the bones.

4 Preheat oven to 400°F (200°C). Add cream to the casserole juices, then simmer in a pan, stirring, until thickened. Stir in the tarragon, rabbit meat, salt, and pepper. Spoon into pie dish and cover with pastry. Brush with egg wash. Bake for 20 minutes, or until golden brown.

Rabbit & ham terrine

With this clear, firm, savory jelly comes the statement textures of ham hock and rabbit meat. Selected vegetables and seasonings complete a stunning culinary combination.

Serves 4

Ingredients

1 ham hock (or 2 if not using rabbit), weighing about 2¼ pounds (1 kg)
1 rabbit, about 2¼ pounds (1 kg), cut into 6 pieces
2 carrots, peeled and coarsely chopped
3 sticks celery, chopped
2 small onions, peeled and halved
8 black peppercorns
8 coriander seeds
1 teaspoon (5 ml) white wine vinegar
3 teaspoons (3 leaves) plain gelatin
¼ cup (30 g) capers, rinsed and drained
¼ cup (30 g) chopped French cornichons or pickles
small bunch flat-leaf parsley, chopped

Chef's Tips

Take care not to let the stock boil during step 1 or it will turn murky. Keep the saucepan over low heat, and remove the impurities as they rise to the surface of the liquid.

Method

1   Put the ham and rabbit into a large pot and cover with water. Simmer for 10 minutes, skimming regularly. Add the vegetables and spices. Simmer gently for 3 hours, skimming regularly, until the meat starts to fall away from the bones.

2   Remove meat from the pan and strain liquid into a bowl. Cool, then skim off fat from the surface. Pour stock into a clean pan and boil until reduced by half. Check for saltiness, and stop cooking if the stock gets too salty. Strip the meat from the ham and rabbit bones and reserve in separate bowls. Line a rectangular loaf pan with a double layer of plastic food wrap, leaving it overhanging.

3   Keep 3½ cups (800 ml) stock in the pan and add the vinegar. Soak the gelatin in a little cold water until soft, then add to stock. Heat gently until the gelatin dissolves and starts to thicken. Cool slightly, then stir in the capers, cornichons, and parsley. When liquid starts to set, pour over the bowls of meat. Pack half the rabbit mix into the pan, then add all the ham mix. Top with remaining rabbit. Cover with the excess plastic. Cut some cardboard the same size as the pan, cover in foil, then press onto terrine as a lid. Weigh down with cans and chill overnight. Turn out the terrine and remove plastic. Cut into slices and serve with pickled cauliflower.

Roast venison with oatcake & huckleberry sauce

Venison's best cut deserves classy company—crumbly cranberry oatcakes and a spiced liqueur gravy.

Serves 4

Ingredients

½ stick (60 g) butter
4 venison loin medallions, each about 6 ounces (175 g)
salt and black pepper

For the oatcake
½ stick (60 g) butter
2 teaspoons (10 g) minced, peeled shallot
1½ cups (350 ml) heavy cream
½ cup (75 g) dried cranberries
1 cup (90 g) quick oats
1 tablespoon (15 ml) beaten egg
1 teaspoon (5 g) chopped fresh chives
salt and black pepper

For the huckleberry sauce
¼ stick (30 g) butter
1 cup (100 g) sliced shallots
1 sprig rosemary
1 star anise
1 cinnamon stick
2 venison shank bones, roasted
¼ cup (60 ml) honey
¼ cup packed (45 g) brown sugar
⅓ cup (80 ml) sherry vinegar
½ cup (120 ml) port
¼ cup (60 ml) crème de cassis liqueur
2 cups (475 ml) brown chicken stock
¼ cup (50 g) fresh huckleberries (bilberries), blueberries, or black currants
salt and black pepper

Butcher's Tips

To roast bones, preheat oven to 350°F (180°C). Roast bones in a roasting pan for 1 hour, then deglaze the pan juices with a little water over medium heat.

Method

1   For the oatcake, melt half the butter in a pan. Add shallots and sweat for 1 minute, or until softened, but not browned. Add cream and cranberries and bring to a boil. Simmer for 1 minute. Stir in oats and cook for 1 minute. Remove from the heat. Add egg, chives, salt, and pepper, and stir for 1 minute. Cool completely, then pack the mixture into a round or rectangular mold, or individual molds. Chill until firm.

2   For the sauce, melt butter in a large pan over medium heat. Add the shallots and sweat for 2 minutes. Add herbs, spices, bones, and the pan juices from the roasted bones. Add honey and sugar, and cook until lightly caramelized. Add vinegar and cook for 1 minute. Pour in port and liqueur, then simmer until the sauce is reduced by two-thirds. Add stock and simmer until mixture is thick enough to coat the back of a spoon. Strain sauce, then add berries. Season to taste.

3   Preheat oven to 400°F (200°C). For the roast venison, melt the butter in a sauté pan. Add the venison and sear over high heat until golden. Place meat in a roasting pan. Roast for 4 minutes, or until the internal temperature reads 120°F (49°C) on a meat thermometer. Set aside the meat and cover to keep warm.

4   Carefully turn out the oatcake mixture from the mold. Sauté the oatcake in the remaining butter over low heat until golden.

5   To serve, carve the venison across the grain and serve with the oatcake. Drizzle the sauce on top.

Barding & larding

Although most game is now farm-reared, it still offers less fat than traditional livestock and so needs extra fat when cooking.

Fat delivers both flavor and quality to cooked game and keeps your cut moist and tender. Since game animals are lean, they can become dry unless lavished with extra fat before cooking. Traditionally, there are two ways to add fat: "Barding" is where a layer of fat is attached to the outside of the meat, while "larding" places the fat right inside the meat.

LARDING

Thin strips of white, firm, dry fat (usually pork) are placed inside the meat using a larding needle, also known as a larder or lardoir, measuring 6–12 inches (15–30 cm) long. The needle is actually a thick, pointed skewer with a short handle. The body of the needle is hollow and open on one side, so it acts as a trough. The strip of fat is placed in the trough, then the needle is pushed all the way through the meat along the grain. The fat stays inside the meat as the needle is drawn back out on the other side. This process is repeated every couple of inches along the cut to ensure an even internal basting. The use of a larding needle is more effective than inserting pieces of fat into the surface of the meat by slitting it with a knife.

Larding is a time-consuming process, but with an expensive, cut, such as a haunch of venison, beef filet or game bird, it is worth doing to get the best from the meat. Your butcher can sell you larding fat, or prepare your meat ready for braising, pot-roasting, or roasting.

BARDING

Barding is an easy way to add fat—butchers do all the time before selling lean cuts. A single layer of firm fat melts and trickles over the meat as it roasts, basting continuously. Fresh pork fatback is ideal because it has a neutral taste and does not dominate the flavor of game, but salt pork and bacon can also be used. Game is not the only meat that benefits from

Serving Etiquette

Do not cook game past medium-rare, as further cooking will result in tough meat. In French cuisine, game is usually served with the barding fat in place, but removed for beef, pork, and poultry dishes.

added fat: Lean beef silverside and topside are often sold wrapped in a snug layer of extra fat, while chicken and turkey breast benefit from barding to prevent dryness—see below.

When barding, you can also spread the meat with butter—seasoned with fresh herbs, if desired—before applying the barding fat. Tie the fat on with kitchen string or use short skewers to secure. If any part of the meat is not covered by the fat, rub with soft butter before roasting.

▲ Bacon-barded chicken stays moist when cooking.

▶ On a lean venison loin, short skewers help keep the barding fat firmly in place.

Venison burgers with root remoulade

Attention fast-food addicts: Try this upmarket version of burger, slaw, and fries—you might never go back to beef.

Makes 4 burgers

Ingredients

1 tablespoon (15 g) butter
3 tablespoons (45 ml) vegetable oil
2 shallots, peeled and finely diced
2 cloves garlic, peeled and minced
1 tablespoon (10 g) chopped fresh thyme
1¼ pounds (500 g) ground venison
4 ounces (115 g) ground fatty pork
1 teaspoon (5 g) ground cumin
1 teaspoon (5 g) salt
1 teaspoon (5 g) ground black pepper
1 egg, beaten

For the carrot & celeriac remoulade
1 small celery root (celeriac), peeled and grated
or cut into very thin batons
4 large carrots, peeled and grated
juice of 1 lemon
½ cup (120 ml) mayonnaise
2 teaspoons (10 ml) Dijon mustard
1 small bunch fresh parsley, chopped
salt and black pepper

For the triple-cooked fries
salt and black pepper
2¼ pounds (1 kg) baking potatoes, such as Idaho
or Maris Piper, peeled and cut into ¾-inch (1-cm)
fries, soaked in cold water
vegetable oil, to fry

To serve
burger buns, shredded lettuce, and sliced pickles

Method

1. Heat the butter and 1 tablespoon of the oil in frying pan over medium heat. Add the shallots and garlic, and sweat for 2 minutes. Add the thyme and cook until the shallots are soft. Allow to cool.

2. In a large bowl, combine both types of ground meat. Wear latex gloves if available, or use wet hands to mix the meats for 2–3 minutes. Add the remaining ingredients and combine thoroughly. Mold into four patties. Chill until needed.

3. For the remoulade, combine all of the ingredients in a bowl. Chill for a minimum of 2 hours to let the flavors develop.

4. For the fries, bring a large pan of salted water to a boil. Add potatoes and simmer for 5 minutes, or until just tender. Carefully transfer potatoes onto a clean cloth. Allow to cool, then refrigerate.

5. In a heavy pan, heat oil to 250°F (130°C). Add potatoes and fry until just crisp to the touch. Remove from the pan, cool, and refrigerate.

6. To finish, preheat oven to 400°F (200°C). Heat the remaining 2 tablespoons oil in a frying pan over medium heat. When smoking hot, add the burgers and fry for 2 minutes on each side, or until browned. Transfer to a baking pan and bake for 6 minutes. Allow to rest, covered, for 2 minutes.

7. Heat the frying oil in a deep-fat fryer or heavy pan to 350°F (180°C). Add the fries and cook for 3-4 minutes, or until golden brown. Drain on paper towels. Season with salt and pepper. Toast the burger buns, then stack with lettuce, burgers, and pickles. Serve with the fries and remoulade.

Wild boar, guinea fowl & apple meatballs

Be brave with your game and combine two contrasting meats for these bold meatballs. Serve with mashed potatoes and slow-cooked red cabbage.

Serves 6–8

Ingredients

½ stick (60 g) butter
2 large onions, peeled and finely diced
4 cloves garlic, peeled and smashed
1 tablespoon (15 g) fennel seeds
8 sage leaves, finely chopped
4 apples, peeled, cored and finely diced
1 cup (60 g) fresh white bread crumbs, soaked in
    ½ cup (120 ml) milk and ½ cup (120 ml) olive oil
2¼ pounds (1 kg) ground wild boar
2¼ pounds (1 kg) ground guinea fowl or other meat
1 egg, beaten
salt and black pepper
2 tablespoons (30 ml) vegetable oil, plus extra for
    greasing roasting pan
⅔ cup (150 ml) hard (dry) apple cider

Chef's Tips

For a thicker sauce, melt 1 tablespoon butter in a pan with 1 tablespoon all-purpose flour. Cook for 1 minute, then whisk in the pan juices. Season well. Simmer, stirring, for 5 minutes, until thickened.

Method

1   Put the butter, onions, garlic, fennel, and sage into a large heavy saucepan. Sweat for 10 minutes, or until the onions are soft. Add the apple, and cook for another 5 minutes. Allow to cool.

2   Add the soaked bread crumbs, ground meats, egg, and seasonings to the onion mix. Add extra bread crumbs if the mixture is sloppy—it should be firm enough to roll into balls. Chill.

3   Grease a Dutch oven or deep roasting pan. Preheat oven to 325°F (170°C). Using wet hands, shape the meat mixture into 14 meatballs. Place in the Dutch oven or pan. Pour in the cider. Cover and bake for 1 hour, adding a little extra cider if the meatballs look dry.

4   Pour the cooking juices into a saucepan and boil until thickened. Heat the remaining 2 tablespoons of oil in a frying pan. Add the meatballs and fry for 2 minutes, or until browned all over. Serve with the cooking juices.

Game casserole

Make the most of the game season with this warming casserole.
A sweet, wine-based gravy brings out the best in the wild meat.

Serves 4

Ingredients

¼ cup (60 ml) vegetable oil
2¼ pounds (1 kg) mixed game, cut into large pieces
½ cup (115 g) chopped smoked bacon (lardons)
¼ stick (30 g) butter
2 onions, peeled and coarsely chopped
1 tablespoon (15 g) dark brown (muscovado) sugar
2 tablespoons (30 g) all-purpose flour
½ cup (125 ml) port wine
2 cups (500 ml) hot chicken or beef stock
2 carrots, peeled and cut into chunks
1½–2 cups (about 150 g) sliced button mushrooms
2 teaspoons (10 g) dried mixed herbs
salt and black pepper

Butchers's Tips

Rabbit, pigeon, pheasant, partridge, and venison all work well in this recipe. Just ask your butcher to select a balanced mix of prepared meat. If using boned meat, allow 6–8 ounces (175–225 g) per person.

Method

1 Preheat the oven to 350°F (170°C). Heat the oil in a heavy frying pan. Add the meat in batches and sear over high heat until browned all over. Transfer to a Dutch oven or a casserole dish. Add the bacon to the pan and cook until evenly browned. Add to the game.

2 Add the butter to the frying pan. Stir in the onions and fry over high heat for 5–10 minutes, until softened and beginning to turn golden. Stir in the sugar. Sprinkle in the flour and cook for 1 minute. Gradually blend in the port and stock. Bring to a boil, stirring constantly. Pour over the browned meat and add the carrots, mushrooms, and herbs. Stir gently to combine.

3 Cover the casserole and bake for 1½ hours, or until the meat and vegetables are tender. Season to taste and, if the gravy is a little too thick, thin it down with extra stock or water.

Buffalo steaks with parsley

Leaner and healthier than beef but just as flavorful, buffalo
should not be cooked past medium-rare or it will be tough.

Serves 4

Ingredients

4 buffalo steaks (New York strip or top sirloin),
3–4 ounces (75 g–115 g) each
1 teaspoon (5 ml) corn oil

For the marinade
¼ cup (60 ml) balsamic vinegar
⅓ cup (80 ml) olive oil
4 cloves garlic, peeled and minced
1 shallot, peeled and minced
¼ teaspoon salt
black pepper

For the parsley butter
¼ stick (30 g) butter, softened
1 teaspoon (5 ml) honey
1 tablespoon (10 g) chopped fresh parsley
¼ teaspoon cayenne pepper, or to taste

Method

1 Place the steaks in a sealable plastic food bag.
Mix the marinade ingredients together in a bowl,
then pour into the bag with the steaks. Seal the
bag and massage the meat through the plastic to
work the marinade into the flesh. Refrigerate for
2 hours, or overnight.

2 When ready to cook, preheat the broiler or grill to
high. Grease the grill rack with corn oil. Remove
the steaks from the marinade, discarding the
marinade. Cook for 4–6 minutes on each side.
Allow the meat to rest under tented foil for
5 minutes before serving.

3 While the meat is resting, mix together the butter,
honey, parsley, and cayenne pepper. Shape into
separate servings and serve on top of the steaks.

Butchers's Tips

You can buy buffalo and elk in the same cuts as
beef, but since these meats have less fat they cook
about 30 percent faster. So if substituting buffalo for
beef in a recipe, reduce the cooking time by a third.

Boar stew with blueberries

Boar is a key player in Tuscan cuisine—this dark, deep-flavored stew is enhanced by fresh blueberries and chocolate.

Serves 4

Ingredients

¼ cup (60 ml) olive oil
2¼ pounds (1 kg) wild boar (any cut), trimmed
  and cut into small cubes
1 onion, peeled and finely chopped
1 carrot, peeled and finely chopped
1 stick celery, finely chopped
1 cup (120 ml) red wine
1 cup (120 ml) pure blueberry juice
2 bay leaves
sprigs of thyme
salt and black pepper
½ ounce (10 g) dark chocolate, finely chopped
6 ounces (175 g) fresh blueberries

Butcher's Tips

Wild boar meat has a sweeter, more intense flavor than domestic pork and is a darker red color. Because it is so lean, take care to add plenty of liquid to the meat to keep it moist. If you cannot find wild boar for this recipe, substitute with a lean cut of pork.

Method

1 Heat a little of the oil in a heavy frying pan over medium heat. Add the meat to the pan in batches and brown on all sides. Add a little more oil if necessary for each batch. Set the meat aside.

2 Heat a little more oil in a Dutch oven or flameproof casserole dish over medium heat. Add the onion, carrot, and celery, and sauté until the onion is golden. Add the wine, blueberry juice, herbs, salt, and pepper.

3 Add the meat, plus enough water to cover it. Bring to a simmer and cook, covered, over low heat for 45–60 minutes, or until the meat is tender.

4 Add the chocolate and stir until melted, then add the blueberries. Cover and cook for 5 minutes. Allow the stew to rest for 10 minutes before serving with potatoes or polenta.

chapter 5
Chicken & Turkey

Whether served as a dressed-up roast or a weekday stir-fry, poultry offers unlimited opportunities for seasonings and presentation. The meat welcomes a world of flavors, and even free-range organic birds will not break the bank.

From Farm to Table

Perhaps more than any other meat, a succulent, crisp-skinned chicken shouts of home and comfort. No wonder it's one of our favorite foods: The smell of a fresh chicken roasting in the hot oven is pure bliss.

Raising Poultry

World chicken consumption increases every year, with nine billion birds born each year in the U.S. alone. But some argue that this popularity comes at a moral cost, and that the meat has an inferior flavor. Although the poultry industry is highly regulated, most chickens are intensively reared in conditions that are questionable in terms of animal welfare.

INTENSIVELY REARED CHICKEN

A "broiler" chicken is the name given to a chicken reared for meat rather than egg production. Broilers experience a brief life placed in a high-density indoor environment, packed into sheds without the room to express natural behavior. To combat the diseases prevalent in intensively reared animals, their feed is supplemented with antibiotics and growth hormones. The incredible success of intensive chicken farming has brought down the cost of the meat because the birds are raised so efficiently and economically. But this success comes at a price—animal welfare and taste are compromised.

FREE-RANGE AND ORGANIC CHOICES

With more awareness of intensive rearing methods, consumers are embracing free-range and organically raised birds. Free-range birds, by law, have access to an outdoor area for some time during the day. In reality, the chicken's experience of the outdoors is limited, but this is a better deal than that of an intensively farmed bird. Organically reared birds enjoy higher welfare standards and a more natural diet. Whichever type of bird you choose, buy from a producer who values the welfare and flavor of their chickens. Your butcher will have selected a supplier based on the quality of the meat.

FLAVOR AND TEXTURE

Only a narrow range of breeds are used to rear chickens commercially, and there is little difference in taste, regardless of production methods. Free-range and organic birds are older—sometimes nearly three times older—than their intensively reared counterparts. This means that they are more flavorful, with well-developed muscles from exercise and foraging, not to mention stronger bones for a great-tasting stock. These chickens have a higher proportion of dark meat because they enjoy more exercise than those in cramped quarters. Intensively reared chicken meat has a mild taste, and most consumers do not question that this is its natural flavor. Corn-fed chicken—with it bright yellow skin—has a

Heritage Breeds

Heritage chicken breeds are noted for their lavish plumage as much as their meat and colorful eggs. A whole palette of brown, rust, black, gray, and white feathers abound on these now-rare and exotic birds. New Hampshires and White Jersey Giants are two particularly confident, hardy varieties with meat that is quite unlike regular chicken. Check your local farmer's market for supplies of heritage chicken.

more distinctive taste, and may be intensively raised or free-range. Smaller producers who raise rare heritage breeds offer genuinely different flavor choices.

Most American chickens are water-chilled, meaning that the slaughtered chicken is cooled in a communal bath of treated water. In contrast, air-chilled chickens from some organic and free-range farmers are chilled on racks in a cold room and, as a result, do not absorb any cooling water. This means that they lose less weight when cooked. Air-chilled chicken is tastier because the juices are not diluted, and the skin is tighter and firmer when cooked. This method is widely used in Europe.

Classification

The chicken goes under several names depending on its age:

▶ **Poussin** The French name for a young chicken, usually less than 28 days old. Since poussin are so small, they cook quicker than a large roasting bird, and have a delightful taste. An ideal choice for a romantic dinner for two.

▶ **Spring Chicken** Another term for a young bird, usually slightly older and heavier than a poussin.

▶ **Cornish Game Hen** Hybrid breed of young chicken.

▶ **Rooster** (U.S.) or Cockerel (U.K.) A male chicken, the classic base of the famous French casserole coq au vin.

▶ **Capon** Originally used in reference to castrated male chickens—a procedure said to improve the meat's flavor. This practice is now rare, but the term is sometimes employed to describe older male chickens.

NUTRITIOUS FAMILY MEAT

Low cost is not the only reason why chicken is a major part of America's diet. The meat is packed with iron, B vitamins, minerals, and protein; it is low in fat when served without skin. Chicken is easy to portion and cook, marrying well with a host of flavorings, from apricots and honey, to spicy chiles. The meat happily accepts hot cooking in a wok or on the grill.

However, clever cooks know that raw poultry can grow harmful bacteria, such as salmonella. Keep raw poultry in a sealed container, at the base of the refrigerator, to avoid meat liquids dripping onto other foods. Wash your hands and (non-wood) utensils thoroughly in hot, soapy water before and after handling poultry. Always thaw frozen poultry before cooking.

Organic chickens receive certified organic feed, and are grown without artificial fertilizers or pesticides. The chickens do not receive hormones or antibiotics and have free access to outdoor pasture and foraging.

Chicken Cuts

There are no limits when you've got a chicken in the kitchen because every cut can be cooked in any way you please. Separate parts make for quicker cooking and easier serving than whole birds.

For such a small animal, cooks are spoiled for choice at the fresh meat counter when buying chicken, and most cuts are available with or without the skin and bone. Whatever cut you choose, do remember that poultry is only safe to eat when it is thoroughly cooked through. Traditionally, cooks and chefs are taught to check that the internal juices run clear from the bird and that none of the meat or bone is pink. Although the flesh on some breeds of chicken remains pink even when it's done, you're better to be safe than sorry on this point. The definite check for doneness is that your meat thermometer registers 165°F (75°C) when inserted at the thickest part of the meat, or at the thigh on a whole chicken.

Drumstick

Taken from the lower part of the chicken's leg, the meat on the drumstick is slightly darker than at the thigh. A perennial grilling and frying favorite, drumsticks take well to strong seasonings, crispy coatings, and sticky glazes. The skin adds essential fat and flavor when roasting, but can be easily peeled off before or after cooking to cut the calories.

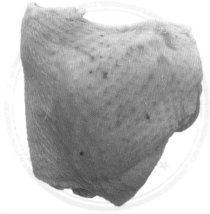

Thigh

The thigh is the portion of the leg above the knee joint. Its pleasantly even shape makes it easy to handle, cook, serve, and eat! Can be skinned for leanness, or left intact for crispy roasting. Also sold boned for stuffing or slicing for frying. Thigh, with its dark, flavorful meat and always-moist texture, is an excellent alternative to breast.

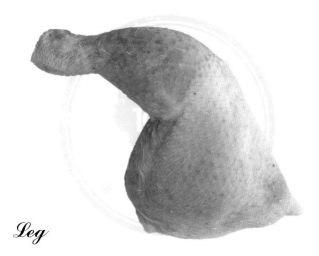

Leg

A whole chicken leg consists of the thigh and the drumstick. It is a great dark-meat cut for roasting and braising—you can slip herbs or spices easily under the skin before cooking to impart extra flavor. Ideal as an economical single serving—your butcher can skin and bone the leg to make it leaner. Or, cook with the skin and skim the fat after.

Wing

The whole wing offers all-white meat and is composed of three distinct sections. The first section is the "drumette," and includes the part between the shoulder and the elbow. The second section, labeled the "mid section," is the flat center of the wing. The third part of the wing is the tip. The mid section may be sold with or without the tip.

A.K.A. Wing flat, mid joint (mid-section sold without the tip)

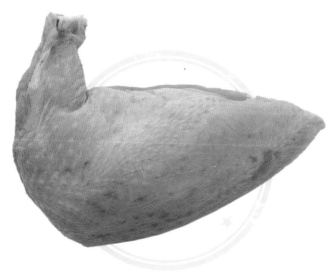

Supreme

A chicken supreme is the breast and wing prepared with, or without, the bone and skin. Cooking on the bone, with the skin, maintains the natural flavor of the meat and keeps it moist. Boneless supremes can be opened, stuffed, then rolled before cooking. The breast and wing can be divided into smaller skinless, boneless cuts that are sold under various names, according to usage. These include chicken tenders, filets, tenderloin, fingers, and strips.

167

Turkey

More than 200 million turkeys are eaten annually in the U.S., mostly at Thanksgiving. Turkey is a thrifty cook's standby, and roast-turkey leftovers offer no end of nutritious family meals.

INTENSIVE VERSUS FREE-RANGE

Since weight gain is often the most important factor for the farmer, most turkeys are fed rich protein foods for rapid growth. In contrast, free-range and organically reared turkeys receive a natural diet and grow more slowly. Try to buy your turkey with its neck and giblets, as these are essential for your gravy. Avoid frozen turkey if you can because the texture and flavor is far inferior to fresh. Turkeys are not only sold whole, but also in parts. You can find legs, breasts (with and without skin or bones), thighs, white meat cutlets, and ground (minced) meat. A turkey crown describes the breast and legs, but no wings. Butterfly breasts are boned and split ready for the grill.

The turkey's natural meat juices combine with additional seasonings to create a wonderfully aromatic basting liquor, and the basis for your gravy.

Perfect Roast Turkey

The classic way with turkey is to roast it whole. For best results, here are a few essential butcher's tips to ensure perfection.

- **DON'T STUFF THE TURKEY'S NECK** because this reduces the temperature of the meat and increases the cooking time.

- **SEASON WELL** inside and out, with salt, herbs, and spices. Oil or butter the meat all over and under the breast skin.

- **DON'T TRUSS**: Turkeys used go into the oven with their legs tied (trussed) to ensure a tidy finish. However, trussing is not always the best idea because it prevents the heat flowing evenly around the meat. If you prefer to truss your turkey, don't tie the legs too tightly.

- **ROAST UPSIDE-DOWN,** then turn it over an hour before the end of roasting to let the breast brown. The breast will be a little misshapen, but it stays really moist. Also, the turkey cooks a lot faster this way and solves the problem of breast dryness.

- **BASTE** regularly for the last hour of roasting to create a golden glow on the skin. Or, use a roasting bag for the main part of cooking, then remove when ready to crisp up the skin.

- **TO CHECK** the turkey is cooked, insert a meat thermometer into the thickest part of the thigh: It should read 165°F (75°C).

- **REST THE TURKEY** after cooking on a warmed platter for 15–20 minutes to let the tasty juices run back into the bird rather than be lost on your carving board as you slice. Keep the turkey covered with foil.

BRINING AND SALTING

By soaking the turkey in salted, seasoned water—known as brining—you can prevent meat dryness and reduce moisture loss by up to 40 percent. However, the process is tricky with a whole bird because of its size and the fact that it needs to be refrigerated during this process. Brining can rob the meat of flavor by diluting the juices and, for some cooks, the meat can taste *too* succulent or watery. For others, the right brining mix imparts essential seasonings and achieves just the right texture.

Dry-salting a turkey lets the meat create its own mix of concentrated brine, but slightly cures it in the process. The finished meat is extremely succulent, but may not suit all palates. Many chefs avoid brining altogether—they just season the meat well and take care not to let it dry out.

Bronze Turkey

Bronze turkey is a cross between wild and domestic turkeys, with elegant, dark plumage. Bronze turkey was once America's most popular breed, but is now a highly priced heritage breed. A free-range bronze turkey takes six months to fully mature, in contrast to intensively reared birds slaughtered at 12 weeks. Reserve your bronze turkey well in advance for Thanksgiving: It will be worth the price. If you like a wild, gamy flavor, try genuine wild turkey instead.

Chicken breasts with beans & tomatoes

White beans add an elegant rusticity to this simple yet satisfying dish, which calls only for fresh, crusty bread to mop up the juices.

Serves 4

Ingredients

4 bone-in chicken breasts, with skin
salt and black pepper
2 cans cooked white beans (such as cannellini or
    Great Northern), weighing 15 ounces (425 g) each
⅓ cup (80 ml) olive oil
½ teaspoon red pepper flakes
12 ounces (350 g) cherry tomatoes, halved
½ cup (120 ml) white wine
8 cloves garlic, peeled and chopped
4 sprigs thyme

Method

1 Preheat oven to 450°F (230°C). Position a rack in the center of the oven. Wash the chicken breasts and pat them dry.

2 Season both sides of the chicken with salt and pepper. Arrange the chicken, skin-side up, in a large casserole dish or Dutch oven.

3 Drain, rinse, and dry the beans. In a bowl, toss the beans with the olive oil, red pepper flakes, tomatoes, white wine, garlic, and thyme sprigs. Season with salt and pepper. Scatter the mixture around the chicken. Drizzle with a little olive oil.

4 Bake for 50 minutes, or until the chicken is cooked through. Remove the thyme sprigs and serve with crusty bread.

Butter-roasted chicken supremes

Herbed butter helps roast this chicken to a crisp and golden finish. Lemon and garlic are the base of the flavoring—just choose your favorite fresh herb to complete the seasoning.

Serves 4

Ingredients

4 bone-in chicken breasts with drumette (chicken supremes), with skin
1 shallot, peeled and sliced
2 tablespoons (30 g) garlic, chopped
salt and black pepper
¼ stick (30 g) unsalted butter

For the herbed butter

¼ stick (30 g) unsalted butter
1 teaspoon (5 g) chopped garlic
2 teaspoons (10 g) finely grated lemon zest
3 tablespoons (30 g) chopped fresh herbs

Butcher's Tips

Some chefs prefer skinless supremes, but this recipe calls for skin-on breasts to hold the butter. Supremes should be sold with the wing (drumette) still attached, but it's wise to check when buying. On the East Coast, supremes are called "Statler breasts," after the historic Boston Statler hotel.

Method

1 For the herbed butter, allow the butter to soften to room temperature. Add the garlic, lemon zest, and herbs, and mash together with a fork. Wrap the butter tightly in plastic food wrap. Chill until ready to use.

2 Preheat the oven to 400°F (200°C). Lift the skin of each piece of chicken slightly and spread the herbed butter underneath, between the meat and the skin, pulling the skin back over when finished.

3 Melt the plain butter in a frying pan or skillet over medium heat. Add the shallot and garlic and sauté until softened. Place the chicken, skin-side down, in the pan. Cook just until the skin is golden brown. Remove the chicken from the pan and place, skin-side up, in a roasting pan. Roast for 20–25 minutes, basting once or twice, until the chicken is cooked through.

171

Sticky chicken wings

Spicy, yet sweet, these sauce-coated chicken wings are perfect for casual get-togethers and great for kids. For convenience, prepare a day or two ahead of time, then reheat to serve.

Serves 4

Ingredients

18–20 chicken wings
2 tablespoons (30 ml) dark soy sauce
½ cup (120 ml) water

For the rub

1 teaspoon (5 g) smoked paprika
1 teaspoon (5 g) ground cumin
1 teaspoon (5 g) celery seed powder
1 teaspoon (5 g) onion powder
1 teaspoon (5 g) black pepper
2 bay leaves, crumbled
4 cloves garlic, peeled and finely diced

For the sticky sauce

1¼ cup (300 ml) tomato ketchup
¾ cup packed (150 g) brown sugar
½ cup (120 ml) clear honey
1 teaspoon (5 g) five spice powder
1 teaspoon (5 g) ground star anise

Method

1 Wash the chicken wings and pat dry. In a large plastic food bag, mix together the rub ingredients. Add the wings to the bag and seal. Shake the bag gently so that the chicken is evenly coated in the spices. Place the bag in the fridge and leave for 24 hours to marinate.

2 Preheat the oven to 350°F (180°C). Remove the wings from the fridge and empty into a large bowl. Add the soy sauce and water and stir gently to coat the wings. Line a baking pan with foil. Spread out the wings in the pan and cover with foil. Roast for 30–45 minutes, turning the wings twice, until the chicken is cooked through.

3 While the wings are roasting, combine the sauce ingredients in a bowl. Remove the foil from over the chicken. Carefully pour or spoon off the excess liquid. Turn up the oven heat to 425°F (220°C). Coat the wings in the sauce and return to the oven for 5–10 minutes, or until the sauce caramelizes.

Leg of chicken "crépinette"

Wrapped in natural caul fat, these stuffed chicken parcels can be served piping hot from of the oven, or chilled and enjoyed picnic-style the next day with mustard, cornichons, and salad.

Serves 4

Ingredients

4 pieces caul fat, measuring 8×12 inches (20×30 cm)
4 boneless chicken legs with thigh attached,
  with skin
salt and black pepper
¼ stick (30 g) butter

Butcher's Tips

The term "caul fat" describes the omentum, a thin, weblike membrane of fat that surrounds the internal organs of certain animals, including pigs, sheep, and cows. Meat or sausages roasted in caul fat are moist and flavorful, as the fat renders during roasting. Pig caul fat is available from reputable butchers and can also be found at Asian markets.

For the stuffing

14 ounces (400 g) chicken breast, diced and chilled
3 ounces (85 g) chicken liver
1 ounce (25 g) pork fat or slab bacon
½ stick (60 g) butter
15 cremini (small brown) mushrooms, sliced
5 shallots, peeled and sliced
½ cup (25 g) chopped fresh sage
1 cup (240 ml) Madeira wine
½ cup (75 g) cubed white bread, crusts removed
¼ cup (60 ml) heavy cream
2 tablespoons (30 ml) brandy
2 cloves
2 bay leaves
1 egg
salt and black pepper

Method

1 For the stuffing, put the chicken breast meat, liver, and pork fat into a bowl and place in the freezer. Leave until the mixture is semi-frozen.

2 Melt the butter in a pan. Add the mushrooms and shallots, and sauté for 4–5 minutes, or until softened. Add the sage and Madeira wine, and simmer until the liquid reduces and the mixture is almost dry. Cool, then place in the fridge to chill.

3 Put the bread and cream into a bowl and set aside to soak. Warm the brandy in a pan, then add the cloves and bay leaves. Steep for 5 minutes, then remove the cloves and bay leaves.

4 Remove the items that are chilling in the freezer and fridge, and combine with all the ingredients, including the egg, salt, and pepper. Chill once more for 30 minutes. Transfer to a food processor and process for 2 minutes, or until smooth.

5 To stuff the chicken, spread out the caul fat on a large sheet of plastic food wrap. Season the chicken legs on both sides with salt and pepper, then place on the fat, skin-side down.

6 Spoon the stuffing into the chicken legs. Wrap the fat around the chicken to secure the stuffing, then wrap again in plastic wrap, twisting at each end to form a tight package. Chill for 30 minutes. Preheat oven to 325°F (160°C).

7 Remove the plastic wrap from the chilled chicken. Melt the butter in a frying pan or skillet. Add the chicken and sear until golden.

8 Transfer the chicken pieces to an uncovered casserole or Dutch oven. Bake for 20–30 minutes, or until the chicken is cooked through. Allow to rest in a warm place for 10–15 minutes before slicing to serve.

Chicken legs with wine & olives

Cooking with wine deepens the flavor of dark chicken meat, and the olives add a piquant tang. This dish can be paired equally well with couscous or with linguine cooked al dente.

Serves 4

Ingredients

4 chicken legs, with skin
3 tablespoons (45 ml) olive oil
6 sprigs fresh thyme or 1½ teaspoons dried thyme
1 tablespoon (10 g) salt
½ teaspoon red pepper flakes
10 cloves garlic, smashed and peeled
1 cup (240 ml) white wine
16 olives, green or black, pitted or whole

Chef's Tips

For a more hearty dish, replace the olives with bacon and onion. Cut 3 slices of bacon into 1-inch (2.5-cm) pieces. Peel and slice 1 large red onion into wedges. Scatter the bacon and onion around the chicken with the garlic in Step 2. Cook as instructed.

Method

1 Preheat oven to 450°F (230°C). Position a rack in the center of the oven. Mix the olive oil, thyme, salt, and red pepper flakes, then rub the mixture into the chicken.

2 Arrange the chicken, skin-side up, in a baking pan or ovenproof dish. Scatter the garlic around the chicken. Roast for 20 minutes, or until the chicken skin begins to brown.

3 Add the wine and the olives to the chicken. Roast for another 25–35 minutes, or until the chicken is cooked through. Let the chicken rest, covered, in a warm place for 10 minutes before serving.

Golden crumbed chicken with cheese

Stuffing chicken with your favorite cheese keeps the meat moist and tender inside, and allows the skin to crisp up to an appetizing golden brown.

Serves 4

Ingredients

8 boneless chicken thighs, with skin
1 cup (125 g) fresh bread crumbs
1 cup (225 g) grated or crumbled cheese, such as Cheddar, Stilton, or Gouda
1 large egg, lightly beaten
¼ cup (25 g) fresh parsley, finely chopped
salt and black pepper
¼ stick (30 g) butter

Chef's Tips

You can also stuff chicken with your favorite sausage meat. Raw pork sausage meat with tarragon and fennel is a hearty choice, as is finely chopped blood (black) pudding or German sausage.

Method

1 Preheat the oven to 450°F (230°C). In a bowl, combine the bread crumbs, cheese, egg, and parsley. Set aside. Place each chicken thigh between 2 sheets of waxed paper and flatten them with a rolling pin or meat mallet.

2 Season the chicken with salt and pepper and place them skin-side down. Scoop a portion of the cheese mixture onto each thigh. Fold the sides of the thigh over the stuffing so that it is tightly wrapped. Tie with kitchen string to secure.

3 Place the stuffed chicken thighs, skin-side up, in a roasting pan or casserole dish. Season with salt and pepper and place a dab of butter on each thigh. Roast for 30–35 minutes, basting occasionally, until the skin is golden brown. Serve with roasted vegetables.

Roasted chorizo-stuffed chicken

Add zest to a regular roasted chicken by stuffing with chorizo sausage. Placing the stuffing between the skin and the meat keeps the chicken extra moist and infuses it with smoky flavor.

Serves 4

Ingredients

1 chicken, about 3 pound 5 ounces (1.5 kg)
2 raw chorizo sausages
1 tablespoon (10 g) fresh thyme leaves
1 clove garlic, peeled and minced
finely grated zest of 1 lime
1 lemon, halved
salt

Butcher's Tips

Chorizo varies considerably in style depending on its provenance. European chorizo—typically from Spain or Portugal—is cured and has a smoky red pepper flavor. In the US, uncooked Mexican-style chorizo is typically flavored with ancho chile powder.

Method

1 Preheat oven to 325°F (160°C). Remove the sausage from its casings and dice the meat as finely as you can. Place in a bowl.

2 Mix together the thyme, garlic, and lime zest. Add the sausage meat, and mix thoroughly.

3 Stuff the sausage mixture under the chicken skin so that it sits between the flesh and the skin. Try to stuff evenly under the skin of the bird, including the legs. Rub the outside of the chicken with the lemon and sprinkle with salt.

4 Place the chicken in a roasting pan and cover with foil. Roast for 45 minutes, then remove the foil and roast for another 45 minutes, or until the chicken is cooked through. Allow chicken to rest, covered, in a warm place for 15 minutes. Carve the chicken into quarters and serve garnished with slices of grilled zucchini.

Chicken pie

Packed with vegetables and seasoned with sausages, this creamy pot pie is the epitome of comfort food.

Serves 4

Ingredients

For the filling
2 tablespoons (30 ml) olive oil
1 onion, peeled and chopped
2 cloves garlic, peeled and chopped
2 leeks, washed and sliced
4 skinless, boneless chicken breasts, diced
salt and black pepper
4 breakfast sausages, chopped into ¼-inch
   (6 mm) pieces
1 cup (240 ml) white wine
1 cup (240 ml) chicken stock
1 carrot, peeled and chopped
1 head broccoli, divided into small florets
1 parsnip, peeled and chopped
½ cup (120 ml) heavy cream

For the pastry
1¼ (150 g) cups all-purpose flour, plus extra
   for dusting
¼ teaspoon salt
1 stick (120 g) chilled butter, cubed
1–2 tablespoons (15–30 ml) ice-cold water
1–2 tablespoons (15–30 ml) milk

Method

1 For the pastry, sift the flour and salt into a bowl. Rub in the butter with your fingers. Make a well in the center and pour in the cold water, mixing to a firm dough. Wrap dough in plastic wrap and chill.

2 Meanwhile, make the filling. Heat half the oil in a pan. Add the onion, garlic, and leeks. Cover and cook over low heat until softened. Set aside.

3 Heat the remaining oil in a frying pan or skillet. Season the chicken with salt and pepper, then add to the pan, along with the sausage meat. Cook for 5 minutes, or until chicken is browned all over. Add the onion, garlic, leeks, wine, and stock. Simmer until the liquid is reduced. Add the carrot, broccoli, and parsnip.

4 Preheat oven to 400°F (200°C). Roll out the pastry on a floured surface so that it is slightly larger than your pie dish. Add the cream to the filling, then pour mixture into the dish. Cover with the pastry lid, securing to the dish with milk. Pierce a hole in the center of the pastry to allow steam to escape. Brush the top with milk.

5 Bake for 40 minutes, checking regularly that the pastry does not burn.

Boned & rolled chicken

With hassle-free carving, this dish always looks good at the table. Try stuffing with cheese instead of this nut-fruit mix.

Serves 4

Ingredients

1 whole chicken, boned-out
salt and black pepper
olive oil, for drizzling

For the stuffing
½ stick (60 g) butter
1 onion, peeled and finely chopped
2 cups (250 g) fresh bread crumbs
¼ cup (25 ml) cranberry sauce
¼ cup (25 g) dried cranberries
½ cup (50 g) pistachio nuts, coarsely chopped
1 tablespoon (10 g) chopped fresh thyme

Butcher's Tips

Boning-out a chicken means that you can stuff the meat, and the chicken will cook evenly throughout. The task is best done by an expert, so unless you have good knife skills, ask the butcher to bone-out a whole chicken for you. Typically, the technique is to start removing the bones at the back and follow the ribs around to the front.

Method

1 For the stuffing, melt the butter in a frying pan or skillet over medium heat. Add the onion and sauté until softened. Remove from the heat and mix in the bread crumbs, cranberry sauce, cranberries, pistachios, and thyme.

2 Preheat the oven to 375°F (180°C). Place the boned chicken skin-side down. Season with salt and pepper. Arrange the stuffing in the center of the chicken and fold in both sides around it to enclose. Tie the chicken with kitchen string to secure the stuffing inside.

3 Put the chicken into a roasting pan, skin-side up. Drizzle oil over the skin and season with salt and black pepper. Roast for 1 hour, basting the chicken occasionally. When the chicken is cooked through and the skin is golden brown, remove and allow to rest, covered, in a warm place for 15 minutes. Carve into slices to serve.

Roasted rosemary chicken with potatoes

Aromatic garlic and rosemary combine to add the finishing touch to a moist and tender roast chicken. Serve with potato wedges and seasonal greens, such as kale or Swiss chard.

Serves 4

Ingredients

1 oven-ready chicken, about 3 pounds
5 ounces (1.5 kg) in weight
1 shallot, peeled
3–4 sprigs rosemary
salt and black pepper
3 tablespoons (45 ml) sunflower oil
2¼ pounds (1 kg) potatoes
1 lemon
8 cloves garlic, peeled

Method

1 Preheat the oven to 400°F (200°C). Rinse and dry the chicken and place in a roasting pan. Cut the shallot into quarters and put into the main body cavity of the chicken, along with 1 rosemary sprig. Season the chicken with salt and pepper and brush with oil. Cover with foil and roast for 45 minutes.

2 Wash and dry the potatoes, but leave the skin on. Cut the potatoes and lemon into large wedges and place in a bowl with the garlic. Drizzle with the remaining oil and season with salt and pepper. Strip off the leaves from the remaining rosemary sprigs and stir into the mixture.

3 Remove the foil from the chicken and baste with the roasting juices. Arrange the potatoes, lemon, and garlic around the bird. Roast for 45 minutes, or until the chicken is cooked through. Allow the chicken to rest, covered, for 15 minutes before serving. If the potatoes are not ready, return to the oven and roast until tender.

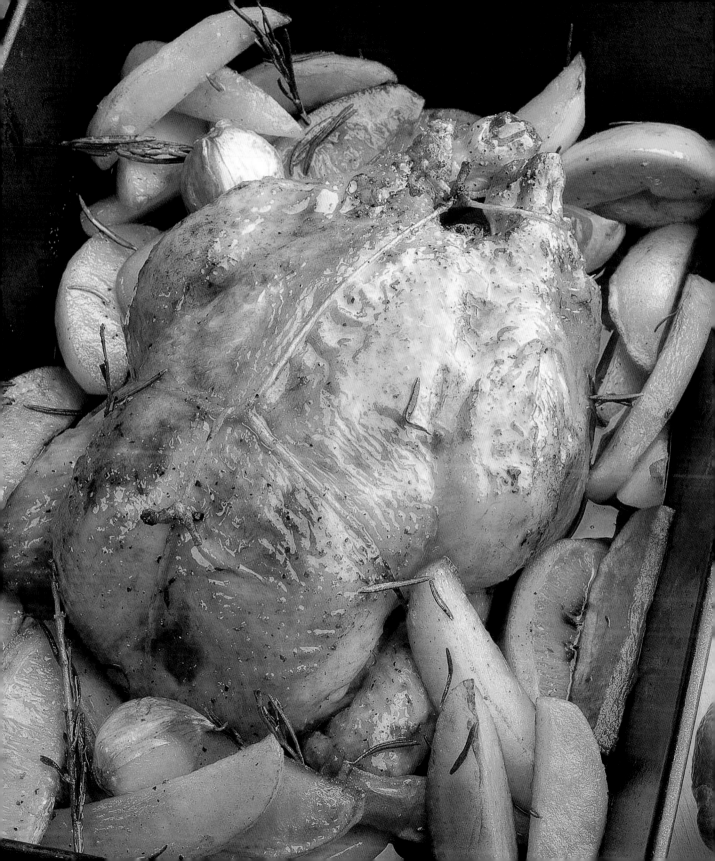

Spatchcocked chicken

Spatchcocking, or butterflying, chicken involves removing the backbone from neck to tail. The bird can be spread flat, allowing it to cook quickly and evenly in the oven or on the grill.

◄ 1. CUT THE SKIN
Place the chicken breast-side down on a work surface. Using a paring knife or kitchen shears, cut along one side of the backbone from the neck to the tail. Repeat on the other side of the backbone; the cuts should be about 1 inch (2.5 cm) apart.

◄ 2. REMOVE THE BACKBONE
Grasp the backbone at the tail end and pull it up and away from the bird. You can discard the bone or use it for making stock. Open up the chicken like a book.

▲ 3. BREAK THE BREASTBONE
Turn over the chicken. Holding it on each side of the breast, pull sharply upward and inward to break the breastbone. Use the knife or shears to carve away the breastbone from the meat. Discard the bone or retain for making stock.

◄ 4. FLATTEN THE CHICKEN
Use the palm of your hand to press down and flatten the chicken. Snip off the wingtips, as they tend to overcook. You can now season the chicken all over with salt and black pepper to prepare for cooking.

Spatchcock roast chicken

Spatchcocked chicken partners well with almost any sauce. Anointing it with a sweet 'n' sour honey-lime marinade gives the skin a crunchy texture and adds a hint of citrus.

Serves 4

Ingredients

3 tablespoons (45 ml) clear honey
juice of 1 lime
1 tablespoon (15 ml) Dijon mustard
1 teaspoon (5 g) ground cumin
salt and black pepper
1 chicken, about 4 pounds (2 kg), spatchcocked

Chef's Tips

To check whether a roasted chicken is cooked through, stick a skewer or sharp knife into the thickest part of the thigh with a spoon placed underneath to collect the liquid—this will look clear when the meat is cooked. The bird's legs will also feel loose if you wiggle them. If using a meat thermometer, the internal temperature should reach 165°F (75°C). This recipe also works well cooked on the grill.

Method

1 In a bowl, mix together the honey, lime juice, mustard, and cumin. Add salt and black pepper, to taste. Brush the mixture over the chicken. Cover and marinate in the fridge for a minimum of 1 hour, or overnight.

2 Preheat the oven to 400°F (200°C). Put the chicken into a roasting pan and roast for 45 minutes, basting occasionally, until the skin is crisp and golden brown and the chicken is cooked through. Let the chicken rest, covered, in a warm place for 5–10 minutes. Carve and serve with a selection of seasonal salads or grilled vegetables.

Deviled chicken livers

The culinary practice of deviling food—preparing it with hot seasonings—dates back hundreds of years. The heat of this fiery appetizer can be tempered by serving with bread or toast.

Serve 4

Ingredients

1 pound (450 g) chicken livers
2 tablespoons (25 g) all-purpose flour
1 tablespoon (15 ml) oil
1 clove garlic, peeled and minced
1 red or green fresh chile, seeded and minced
2 scallions (spring onions), thinly sliced
2 tablespoons (30 ml) white wine
8 ounces (225 g) cherry tomatoes
1 teaspoon smoked paprika
2 bay leaves

Butcher's Tips

If you get your chicken livers from the butcher, they will most likely already have the bile ducts removed. Otherwise, be sure to check for a little green sac and remove it along with any traces of green liquid. This is bile, which will make the livers taste very bitter.

Method

1 Coarsely chop the livers and remove any sinew. Heat the oil in a large frying pan or skillet. Add the liver and cook over high heat until browned.

2 Pour off any excess liquid. Stir in the garlic, chile, and scallions (spring onions). Add the wine, tomatoes, paprika, and bay leaves.

3 Cook over medium heat for 15 minutes, stirring occasionally, until the livers are just cooked in the center and the sauce is reduced. Serve with toasted sourdough bread to soak up the juices.

Lemon poussins

Surrounded by roasted vegetables and cooked in a creamy lemon-infused sauce, these individual mini chickens are sophisticated enough for any special occasion.

Serves 2

Ingredients

2 poussins
salt and black pepper
1 lemon, cut into wedges
4 sprigs tarragon
12 cloves garlic
2 small onions, peeled and diced
1 stick celery, diced
1 carrots, peeled and diced
4–8 new potatoes
1 tablespoon (15 ml) olive oil
½ cup (75 g) canned haricot or other white beans
2 tablespoons (30 ml) white wine
¼ cup (60 ml) chicken stock
¼ cup (60 ml) crème frâiche

Chef's Tips

A poussin is a young chicken aged 4–6 weeks, typically weighing about 1 pound (450 g). Poussin have a light, delicate flavor and not a lot of flesh, so allow one bird per person.

Method

1 Preheat oven to 400°F (200°C). Place 1 lemon wedge and 1 tarragon sprig in each bird. Set aside.

2 Place the garlic, onions, celery, carrots, and potatoes in a roasting pan and drizzle with oil. Roast in the oven for 20 minutes.

3 Remove the pan from the oven and nestle the birds among the vegetables. Add the remaining lemon wedges and season with salt and pepper. Roast for 25 minutes.

4 Reduce oven temperature to 350°F (180°C). Add the beans to the pan. Combine the wine, chicken stock, and crème frâiche, and pour over the vegetables. Roast for another 25 minutes, basting with the pan juices occasionally, until the poussin are cooked through.

Coq au vin

Nothing sums up French country cooking better than this age-old recipe. For a deeper, more provincial flavor, use an older bird or just use the chicken legs for the long, slow braise.

Serves 4

Ingredients

1 tablespoon (15 ml) vegetable oil
8 ounces (125 g) baby onions, peeled
1 onion, peeled and finely chopped
2 cloves garlic, peeled and chopped
1 carrot, peeled and chopped
1 stick celery, chopped
5½ ounces (150 g) pancetta, sliced
3½ ounces (100 g) sliced button mushrooms
1 chicken, cut into parts, or 4 large chicken legs, with skin
salt and black pepper
2–3 sprigs thyme
2¼ cups (500 ml) full-bodied red wine
2 tablespoons (30 ml) brandy, optional
⅔ cup (160 ml) chicken stock

Chef's Tips

The older the bird, the richer the sauce. Tell your butcher you need a chicken for coq au vin, and ask him to order in a mature organic bird specially for you. With access to fresh air, organic birds have the chance to build the stronger, thicker bones that are essential for a sumptuous sauce.

Method

1 Heat the oil in a large pot or Dutch oven over medium heat. Add the onions, garlic, carrot, celery, and pancetta. Cook until the pancetta is crispy, but not burned. Add the mushrooms and cook until slightly colored. Remove the mixture from the pot, but leave the fat behind.

2 Season the chicken pieces with salt and black pepper, then add to the pot with the thyme. Cook until the chicken is golden.

3 Return the vegetables to the pan. Add the wine and brandy, if using. Pour in enough stock to cover the chicken. Partially cover the pot with a lid, then cook over very low heat for 1½–2 hours, until the chicken is very tender. For a thicker sauce, remove the chicken and boil down the cooking liquid until reduced. Return the chicken to the pot to reheat, then serve with steamed potatoes.

Roast turkey with gravy

A small turkey can serve six people and still leave tasty leftovers for the next day. Chefs agree that turkeys taste better if cooked upside-down, then turned for the final browning.

Serves 6–8

Ingredients

1 stick (115 g) unsalted butter
1 turkey with giblets, about 8 pounds (3½ kg)
salt and black pepper
For the gravy
giblets from the turkey, rinsed
1 quart (1 liter) water
1 small onion, peeled and quartered
1 carrot, peeled and chopped
2–3 stalks fresh parsley
2 bay leaves
2 tablespoons (30 g) all-purpose flour
salt and black pepper

Chef's Tips

To truss or not to truss? For many cooks it's a question of preference whether to tie or pin the legs and wings in place. An untrussed bird may cook more evenly (especially in the leg joints) and the amount of skin browning will be maximized. If you do truss your bird, don't tie it up too tightly or you will end up with overcooked breast, uncooked thighs, and pale skin where it is not exposed to the oven heat.

Method

1 Preheat the oven to 350°F (180°C). Fold in the turkey neck flap. Truss the turkey, or leave loose if preferred. Butter a large roasting pan. Place the turkey breast side-down in the pan. (As the breast is not flat, the turkey won't sit straight.) Season turkey with salt and pepper and cover loosely with foil. Roast for 2½ hours.

2 Turn over the turkey to face breast-side up. Spread the butter over the breast and leg area. Roast for 30 minutes, basting occasionally, until turkey juices run clear when the legs are tested with a skewer. Rest, in a warm place, for 45 minutes.

3 Make gravy while the turkey roasts in step 1. Place giblets in a pan and cover with water. Bring to a boil, skimming off the scum from the surface. Add the onion, carrot, parsley, and bay leaves. Simmer for 1½ hours, then strain.

4 When the turkey is cooked, spoon the excess fat from the surface of the juices into a saucepan. Mix in the flour. Cook, stirring, for 1 minute. Gradually whisk in the stock over low heat. Season to taste with salt and pepper, then continue stirring until the gravy thickens.

191

chapter 6
Duck, Goose & Game Birds

Feathered game offers a tasty alternative to regular poultry and a little of this precious meat goes a long way. Take care to select birds with clear, soft skin, and without bruising, blemishing, or tears. Enjoy with a good red wine. Bon appetit!

From Farm to Table

You can order duck, goose, and game birds from your butcher
or local game preserve, then cook with hearty enthusiasm.
And if you are lucky enough to have a hunter to cook for,
you'll be eating some amazing wild birds, too.

Although most of our duck and goose meat comes from intensively farmed birds, their history as hunted birds means that in the kitchen they can be treated as game, rather than as poultry. Despite being waterfowl, duck and goose can be reared without any access to water.

Duck and goose are also reared for their feathers, down, and eggs. Connoisseurs adore duck eggs. With their creamy yolks, they are richer than chicken eggs and ideal for baking. Since duck eggs are produced on a smaller scale than chicken eggs, always check they are fresh before adding to any mixture, such as cake batter. Simply break each egg into a separate bowl before using, and discard any with blood or a spoiled aroma. To avoid the risk of food poisoning, do not eat raw duck eggs.

SUPER FAT

Both duck and goose have wonderful swathes of fat on the breast; this fat is thickest on farmed meat because the birds are less active. The fat is designed to keep the bird both warm, dry, and buoyant in water. It is not marbled in the meat, but sits separately on the flesh. The fat dissolves during cooking, a process called rendering. The fat is a valuable ingredient and a wonderful medium for frying potatoes and other vegetables.

WHITE OR DARK?

Because duck and geese are birds of flight, their breast meat is darker than chicken and turkey. This is because more oxygen is needed by the working muscles, and the oxygen is delivered to those muscles by the red cells in the blood. One of the proteins in meat, myoglobin, holds the oxygen in the muscle, and gives the meat its distinct dark color.

Duck and Goose

Ducks are fed corn and soybean, fortified with vitamins and minerals. The feed contains no animal by-products or hormones, and antibiotic use is very rare. In terms of meat purity, intensively raised duck is healthier than the equivalent chicken or turkey. Additives, preservatives, MSG, and colorings maybe added to processed duck products, such as pâté or smoked breast meat. Duck sold to the mainstream market is classified as Grade A quality—plump, meaty birds with tear- and bruise-free skin. There are no feathers, broken bones, or missing parts. Grade B and Grade C ducks are not stocked by reputable retailers.

Hanging Game Birds

Butchers and hunters spend many a relaxed hour discussing ideal aging times for game birds. Aging lets the meat enzymes break down and tenderize the flesh. Wild pheasant, grouse, partridges, and geese should be hung for three–seven days, depending on the size and age of the bird. Pen-raised pheasants, all quail, woodcock, snipe, and wild ducks need only one–three days. Smaller, younger birds need the least aging. The ideal hanging temperature is 104°–122°F (40°–50°C), depending on whether the bird is intact or gutted and cleaned. Do ask your butcher how the meat has been handled prior to sale.

Geese can distinguish everyday noises from other sounds. As such, they are good as "watch animals." They were used by the Romans to detect the enemies' approach, and during the Vietnam war, U.S. soldiers used geese to warn them of enemy infiltration.

LIVING A BETTER LIFE

Organic duck and goose offers a stronger flavor because the combination of grain and naturally foraged food produces a slightly different meat. Check that your free-range bird has been "pasture raised," with access to water and foraging. To ensure succulent, tender meat, organic and free-range ducks are slaughtered between 10 and 14 weeks of age.

Most duck meat is derived from mallard-related breeds. White Pekins are the basis of the American duck industry, although the breed was only introduced in 1873. Heritage breeds include Aylesbury, Rouen, Khaki Campbell, and Nantes. Mallards are also reared for sports—ask your local shoot preserve about sourcing freshly shot birds.

The only duck breed that is not descended from the mallard is the Muscovy, which originates in South America. This bird is more tranquil than other varieties and "coos," rather than "quacks." Slow to mature, the meat has 50 percent more lean breast meat and 30 percent less fat than other breeds.

GOOSE

The domestic goose, bred in ancient Egypt, China, and India, arrived in America from Europe, where it is a prized meat. Geese are usually raised indoors for the first six weeks, then put out on the range for 14–20 weeks where they eat grass and grain. California and South Dakota are the main geese-raising states. There are 11 key breeds, the most common being the blue-eyed Embden. Pilgrim geese are known as excellent roasters, while Toulouse are a favorite for fois gras pâté. For best results, order a fresh goose and giblets from your butcher.

Game Birds

Hunting access to wild birds varies according to location and habitat. There are many types of feathered game in upland regions, and also smaller birds in marshy wetlands. The size of birds varies from large, such as wild turkey, to very small, such as quail. Before cooking, wild game birds must be professionally bled, cleaned, and drawn. Wild birds spend so much time flying, that the breast meat is often as dark as the leg. Their size determines the cooking time, and a small bird can be roasted in just ten minutes.

If you are a keen hunter, freeze a batch of small game birds (cleaned and drawn) during the shooting season, then cook up for a feast on Christmas Day, allowing a couple of birds per person. Or, ask your butcher what game he can source and prepare for you. Serve game birds with a gravy made in the roasting pan and a jelly made with cranberries or rose-hips.

Duck & Goose

Famed for its fat, compact meat, and intense flavor, duck and goose have been the favored birds of royalty since Roman times. Remember to save any excess fat because it can be used to flavor other dishes.

Whole duck, duckling, gosling, and goose are sold with or without their giblets and neck. The organ cuts are often discarded or exported, but are ideal for stock and gravy. Bone-in cuts such as leg, breast quarter, and breast are an easy roasting option if you do not need a whole bird.

Boneless duck breast meat, skin-on or skinless, is ideal for quick-cooking recipes. Hot, deep-fried duck tongues and duck feet simmered in spiced chicken broth are exceptionally tasty Chinese delicacies—ask your butcher to source these for you or check out your local Asian market and buy by weight.

Whole Duck

Ducks are classed by age: under 8 weeks is a "broiler" or "fryer." Birds between 8 and 16 weeks are "roasters," and those over 6 months are "mature." Roast the bird over a rack to catch the tasty fat. Use the fat for frying potatoes, vegetables, and making savory sauces.

A.K.A. Teal, widgeon (smaller varieties of wild duck)

Breast

The duck breast comes with a layer of thick skin. To cook, score the skin and season well. Fry, skin-side down, for 5 minutes, then drain fat. Turn and cook the meat for 3 more minutes, covered. Rest the meat for 5–10 minutes, then serve medium-rare. Breasts also roast well in a hot oven.

A.K.A. Magret (bone-in or boneless)

Leg

With more connective tissue than the breast, duck leg is best braised. Once the meat is soft, it can be fried to crisp up the skin, or served as a casserole. The dark, rich meat works well with Asian spices and citrus flavors. Legs from smaller ducks can be fried, skin-side down, over low heat until the skin is brown and the meat is cooked through.

Duck Liver

Duck liver looks clean, dry, and plump, with a matte gloss surface and a sweet smell. Avoid slimy livers or those with a sour odor. The texture of cooked duck liver goes beyond creamy chicken liver, and offers a whole new level of richness. Marry the sautéed livers with the sweet-tart contrast of cooked rhubarb, cherries, and a splash of balsamic vinegar.

How to Roast Goose

Goose legs take longer to cook than the breast. This has been a perpetual conundrum for chefs because, if the bird is cooked whole, the breast meat can easily dry out. To cook goose to perfection, cut off the legs and cook them separately as you would for confit of duck—see page 208 for details. The legs can be cooked earlier in the day, then reheated in the oven and served with the breast. Roast the breast in a medium oven, with the breast facing down. The goose fat will render and melt off into the pan, keeping the breast moist as it cooks. About 15 minutes before the goose is cooked, drain off the fat and reserve, then turn the bird over to burnish the skin. The goose breast should be served slightly pink.

Whole Goose

Geese are larger than ducks, and also possess a thick layer of breast fat and close-textured meat. Younger birds are used for roasting; older birds are excellent braised. Save the superbly flavored excess fat for cooking the vegetables that accompany the meal, or allow to cool and refrigerate for later use.

Game Birds

The diminutive size of these little birds makes them irresistible at the dinner table, and their sublime flavor more than compensates for their modest weight.

If you press the breast of a game bird with your fingertips, its response tells you how to cook it. Soft and pliable, the bird is young enough to roast; hard and firm, it is more suited to slow braising. Your butcher can also tell the bird's age by examining the degree to which its beak and feet have developed and hardened. Elaborate French recipes abound for game birds, but a simple roasting is also sufficient. A hot oven is fine, just as long as you lard the sensitive meat with a few strips of bacon. Fresh game birds have a limited season, but are sold frozen all year round.

Partridge

Partridge is hunted or farmed in preserves. Young birds may be roasted, while older birds suit braising in wine. Other game that can be prepared in this way include ptarmigan, sage hen, prairie chicken, chukar, bobwhite quail, and guinea fowl.

A.K.A. Grouse

Quail

This tiny bird—with its lean, dark, gamy meat—can be cooked like a regular chicken or partridge. For a special meal, ask your butcher to prepare semi-boneless whole quails that are "sleeve boned," leaving only the drummette and wing bones intact. When buying, look for moist flesh with creamy yellow skin. Avoid any birds that look dry or that have an aroma. Buy one bird per person.

198

Treasures From the Shoot

- **RAIL (MARSH HEN AND GALLINULE)** These small, lean marsh birds are ideal for slow braising and are hunted for sport across southeastern states.

- **WOODCOCK** French hunters roast this bird whole with the innards, then serve on toast that is spread with the intestine. Roasting the gutted bird is more palatable.

- **SNIPE** Roast these minuscule marsh birds for 10 minutes in a very hot oven.

- **PIGEON (INCLUDING SQUAB AND DOVE)** Soak in milk before braising or roasting.

- **WILD TURKEY** Small and lean compared to its domestic counterpart. Ensure the bird has been bled and cleaned properly at the shoot.

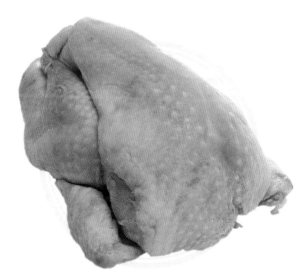

Guinea Fowl

Although guinea fowl are found in the wild—where they forage freely in large flocks—most commercially available birds are organically farm-reared. Native to Africa and popular in French cuisine, this hardy bird offers strong-tasting meat that suits bold flavorings, such as garlic and lemon. Fresh guinea fowl are available in late summer.

Pheasant

For centuries, this magnificently plumed bird has been hand-reared for shooting, but has not lost any of its wild flavor. It suits all cooking methods, and one bird yields two good portions. The bird is aged before plucking, and hens are favored over cocks in terms of flavor. The breast is tender when roasted, but the legs and other cuts suit longer cooking.

199

Warm pigeon salad

This elegant starter of tender pigeon breast, green beans, and seasonal salad greens is a breeze to prepare. Other game birds, including quail and partridge, work equally well in this recipe.

Serves 4

Ingredients

⅓ pound (150 g) green beans
1 tablespoon (15 ml) vegetable oil
4 skinless pigeon breasts
4 slices smoked bacon, chopped
1½ cups (about 100 g) mixed salad greens, to serve
  (arugula, watercress, radicchio, and frisée)
2 small shallots, peeled and thinly sliced
salt and black pepper

For the dressing

2 tablespoons (30 ml) white wine vinegar
1 tablespoon (15 ml) hot English mustard
1 tablespoon (15 ml) whole grain mustard
2 teaspoons (10 g) sugar
⅔ cup (160 ml) vegetable oil

Method

1 Blanch the beans in a pan of boiling water for 1 minute. Drain and set aside. Combine all the dressing ingredients, except the oil, in a bowl. Slowly whisk in the oil until incorporated.

2 Heat 1 tablespoon oil in a frying pan over medium heat until smoking. Add the pigeon breasts and cook for 1½ minutes on both sides. Cover and set aside. Add the chopped bacon to the pan and cook for 2 minutes, or until golden. Add the beans and toss together in the pan for another 30 seconds.

3 Place the salad leaves in a mixing bowl. Gently stir in the sliced shallots, bacon, and beans. Season with salt and black pepper, then drizzle with most of the dressing.

4 Divide the salad among plates. Carefully slice the pigeon breast and lay over salad. Drizzle the reserved dressing on the plate, then serve.

Breast of duck with duck cake & compote

A savory cake, packed with tender duck meat and a delicate mix of root vegetables, contrasts with a fruity compote and sautéed duck breasts. A great prepare-ahead recipe!

Serves 4

Ingredients

4 duck breasts with skin
1 teaspoon (5 g) coarsely cracked black pepper
1 teaspoon (5 g) chopped fresh thyme
1 clove garlic, peeled and crushed

For the duck cake
1 cup (150 g) mixed, finely diced root vegetables
   (carrot, celery root, turnip, and leek)
2 tablespoons (25 g) duck fat
1 baking potato, steamed with the skin removed
8 ounces (225 g) finely diced, cooked duck meat
2 tablespoons (20 g) chopped chives
salt and black pepper
¼ stick (30 g) butter

For the duck compote
¼ cup (60 ml) honey
¼ cup packed (45 g) dark brown sugar
¼ cup (60 ml) sherry vinegar
juice of 1 orange
1 star anise
1 stick cinnamon
¼ cup (50 g) each of dried cherries, golden raisins,
   dried cranberries, and dried blueberries

Chef's Tips

Keep the precious duck fat and store in a sealed jar, in the refrigerator, for up to 2 months. Use when making soups, sauces, roasted meats, and vegetables.

Method

1 Season the duck breasts with the cracked pepper, thyme, and garlic. Marinate for 1 hour.

2 Meanwhile, prepare the duck cake. Gently cook the vegetables in the duck fat until just tender, but not soggy. Grate the cooked potato into a mixing bowl, then gently stir in the diced, cooked duck meat, vegetables, and chives. Season with salt and pepper. Pack the vegetable-duck mixture into a ring mold or rectangular baking pan. Chill for 30 minutes, or until firm.

3 For the compote, heat the honey and sugar together in a heavy saucepan over low heat until bubbling. Add the vinegar, orange juice, and spices. Simmer for 1 minute. Add the dried fruit and simmer for 4–5 minutes, or until the liquid is thick enough to coat the fruit. Set aside.

4 To cook the duck breasts, place the meat in a frying pan over low heat, skin-side down. Cook for 10–12 minutes, or until the fat renders down and the skin begins to crisp. Increase the heat. Sear the meat all over and cook until medium-rare. Remove and keep warm. Reserve the fat.

5 Carefully turn out the chilled duck cake. Melt the butter in a frying pan over medium heat. Add the duck cake and sauté until golden all over. Reheat the compote over low heat.

6 Slice the duck breasts and serve with the duck cake and fruit compote.

Duck with plums

Fresh plums, thyme, and port are a perfect foil for rich roasted duck breasts. This recipe is easy enough for a midweek supper and fancy enough to serve at a dinner party.

Serves 4

Ingredients

1 tablespoon (15 g) juniper berries
1 teaspoon (5 g) fresh thyme
4 Muscovy duck breasts, with skin
salt and black pepper

For the plums

10 sprigs thyme
2 bay leaves
8 plums, halved and pitted
2 teaspoons (10 ml) olive oil
sugar

For the sauce

1 cup (240 ml) port
½ cup (120 ml) balsamic vinegar

Chef's Tips

To crisp the skin, preheat the broiler to maximum. Broil the cooked duck breasts in the roasting pan, skin-side up, for 3–7 minutes, until golden.

Method

1 Preheat the oven to 400°F (200°C). Using a grinder, finely grind the juniper berries and thyme leaves. With a sharp knife, make shallow slashes into the duck skin. Season with salt and pepper, then rub in the juniper mixture. Allow the duck to reach room temperature before cooking.

2 Place the duck breasts, skin-side down, in a roasting pan and bake for 15 minutes. Remove from the oven and pour off fat. Turn the breasts and bake for 15–20 minutes, or until a meat thermometer reaches 135°F (57°C), for medium-rare. Rest the meat, covered, for 10 minutes.

3 For the plums, scatter half the thyme sprigs and bay leaves in a baking dish. Add the plums, skin-side up, and drizzle with oil. Add the remaining thyme and bay leaf. Bake for 20 minutes. Taste the plums and, if too tart, sprinkle with sugar, to taste.

4 Meanwhile, bring the port and vinegar to a boil in a small pan. Reduce the heat and simmer for 5 minutes, or until the liquid is reduced by a half. Thinly slice the duck breasts against the grain and serve with the plums and sauce.

Duck & sarladaise potatoes

Crispy new potatoes fried in duck fat, pungent smoked garlic,
and a fruity sloe gin sauce add a bold French twist to oven-
roasted duck breasts.

Serves 4

Ingredients

For the smoked garlic purée
1 bulb smoked garlic, split in cloves
¼ cup (60 ml olive oil
sprig of rosemary
splash of water
dash of heavy cream
salt and black pepper

For the duck
4 duck breasts, with skin
salt and black pepper

For the sarladaise potatoes
1½ pounds (680 g) boiled new potatoes, cut
　into ¼-inch (6-mm) slices
1 clove garlic, peeled and sliced
handful of chopped fresh parsley

For the sloe gin sauce
¾ cup (200 ml) sloe gin
1 juniper berry, crushed
¾ cup (200 ml) chicken stock

Method

1 Preheat the oven to 325°F (160°C). For the garlic purée, place the garlic cloves, oil, rosemary, and a splash of water in a roasting pan. Cover with foil and roast for 30 minutes, or until the garlic is soft. Squeeze out the garlic flesh into a bowl. Add cream, salt, and pepper, and blend until smooth.

2 For the duck, increase oven 375°F (190°C). Heat a frying pan over medium heat. Score the breast skin and season with salt and pepper. Place skin-side down in the pan and cook for 6–8 minutes, or until the fat renders down and the skin turns golden. Remove the breasts, reserving the fat, and transfer to a roasting pan. Roast, skin-side up, for 7 minutes for rare, or 10–11 minutes for well-done. Rest the meat for 3 minutes.

3 For the sarladaise potatoes, fry the boiled potatoes and garlic in the duck fat until golden brown. Add parsley and salt to taste.

4 For the sauce, boil the sloe gin, juniper, and stock until syrupy and reduced by two-thirds. Season and add any juices from the resting duck breasts. Slice the duck breasts and serve with the hot sauce and potatoes.

205

Duck cassoulet

This rich and filling stew is not difficult to make and makes the most of pantry ingredients. In the hot oven, the individual ingredients combine together beautifully.

Serves 4

Ingredients

4 duck legs, with skin
4 Toulouse or other good-quality pork sausages
½ cup (115 g) chopped smoked bacon (lardons)
1 onion, peeled and chopped
2 carrots, peeled and chopped
2 sticks celery, sliced
6 cloves garlic, peeled and sliced
2 tablespoons (30 ml) tomato paste (purée)
1 cup (225 g) canned chopped tomatoes
salt and black pepper
2 teaspoons (10 g) paprika
½ teaspoon cayenne pepper
2 sprigs fresh thyme
2 bay leaves
14 oz (400 g) canned cannellini, navy, or haricot beans, rinsed and drained

Chef's Tips

One of the great things about this recipe is that it tastes even better the next day. For added contrast, reheat the cooked cassoulet with a layer of fresh bread crumbs to provide a crisp and golden topping. The recipe can also be made with goose. Ask your butcher to order the meat in advance.

Method

1 Preheat oven to 350°F (180°C). Heat a heavy frying pan over high heat. When smoking hot, add the duck legs skin-side down. Cook until the skin is crisp and golden, then turn over and cook the other side until browned. Transfer duck to a Dutch oven or ovenproof casserole dish. Reduce the heat under the pan. Add the sausages and bacon and cook until browned. Cut the sausages in half and add to duck, along with the bacon.

2 Spoon off most of the fat from the pan. Add the onion, carrots, celery, and garlic. Cook, stirring regularly, for 10 minutes, until softened. Add the tomato paste (purée), tomatoes, and 2 cups (475 ml) water. Season with salt and pepper, paprika, and cayenne pepper. Bring to a boil.

3 Pour the vegetable mixture over the meat. Add the thyme and bay leaves. Cover and bake for 1 hour. Add the beans, then bake for another 45 minutes, or until the duck and vegetables are tender. Add a little more water to the cassoulet if it looks dry. Check the seasoning again. Serve with crusty bread, green salad, and a good French red wine.

Confit, pâtés & terrines

Before the invention of refrigeration, cooks prepared confits, terrines, and pâtés as a way to preserve whole cuts and offcuts.

Confit means "preserved" in French, and refers to the preservation of meat with salt and fat. Confits are mostly used to treat duck, but goose, turkey, and pork can also be prepared in this way. The meat is cured in salt and spices, then poached in fat until tender. Spices, such as cloves, juniper, and allspice, or woody herbs, including bay or rosemary, are added to the meat as it cooks. The cooked meat is packed into sterilized jars, then cooled and sealed. The fat sets solid on cooling and encases the meat. Traditionalists claim that meat preserved in this way can last for weeks, or even months, without refrigeration. However, to avoid botulism poisoning, it's best to chill and consume within a few days.

PÂTÉS & TERRINES

Meat pâtés and terrines both use seasoned meats are that cooked, then cooled and served cold. The term "pâté" refers to the pastry case in which the meat was originally cooked, while "terrine" describes the pot which was used as its container. Fat and salt are essential ingredients in all pâté recipes. For pork recipes, the fat is derived from belly and shoulder cuts. For contrasting textures, chefs often combine finely ground meat and a separate mix of diced cuts. Duck, goose, and chicken livers add a uniquely smooth richness that works well on their own or with other meats.

Mousse-like liver pâtés can be packed into molds and garnished with chopped hard-boiled eggs. Game, such as rabbit or guinea fowl, are ideal additions because they have a strong taste that marry well with other flavorings, such as brandy and Madeira wine.

Duck Pâté

Grind together 1 bay leaf, 2 cloves, 8 peppercorns, and 2 allspice berries. Boil 1 finely chopped onion with 6 ounces (175 g) chopped salt pork fatback for 2 minutes, then drain. In a food processor, blend 1 pound (450 g) duck liver with ½ cup (120 ml) heavy cream, 2 eggs, 1½ teaspoons salt, ¼ cup (60 ml) brandy, and the ground spices until smooth. With the processor running, gradually add the pork and onions. Process again for 2 minutes. Sieve mixture through a fine strainer, then pour into a mold or ramekins, cover tightly with foil, and place in a roasting pan. Fill pan with boiling water to just under the edge of the mold. Carefully transfer to the center of the oven. Bake for 1½ hours, or until the pâté is set and registers 160°F (71°C) on a meat thermometer. Cool, then chill the pâté overnight before serving.

COOKING METHOD

Pâtés and terrines are traditionally baked in a bain marie—a roasting pan filled with water. This provides a gentle heat that cooks without drying. Cooked terrines are weighted down as they cool to compress the mixture. An extra layer of calf's foot jelly maybe added as a final seal on the surface.

▶ For duck confit, cured, cooked cuts are packed into jars with salt, herbs, spices, then covered with the original cooking fat.

Super-easy duck ragout with tagliatelle

Gently simmering freshly roasted duck meat with simple seasonings produces a melt-in-the-mouth ragout. The dish is an ideal partner for homemade pasta.

Serves 4

Ingredients

1 duck, about 6 pounds (2.7 kg), quartered
2 onions, peeled and sliced
1 bay leaf
2 cloves garlic, peeled and sliced
salt and black pepper
2 tablespoons (30 ml) tomato paste (purée)

To serve
1¼ pounds (500 g) fresh egg pasta, such as tagliatelle

Butcher's Tips

Duck breasts often cost almost as much as a whole duck, so buying a whole bird will ultimately get you more meat for your money. Your butcher will gladly quarter the bird for you so that you can freeze portions for future use. Use the carcass for making stock and any leftover roasted meat for stir-fries.

Method

1 Preheat oven to 400°F (200°C). Cut the duck into quarters and place in a baking pan. Roast in the oven for 35 minutes, basting occasionally.

2 Drain some of the fat from the roast duck into a large, heavy saucepan. Heat the oil, then add the sliced onions, bay leaf, and garlic. Fry gently over low heat until softened.

3 Remove the skin from the duck and strip the meat from the bones. Cut the meat into small pieces and add to the onion mixture with 1 cup (240 ml) water. Season well with salt and pepper, and stir in the tomato paste. Cover and cook gently for 1 hour, or until you can see the meat almost melting into the sauce.

4 Boil the tagliatelle in a large pan of water for 2–5 minutes, or until just cooked. Drain, then serve with the ragout.

Pheasant Wellington & red cabbage

Encased in golden puff pastry, pheasant breast and delicate button mushrooms remain both succulent and tasty. Serve with velvety sweet red cabbage and mustardy mashed potatoes.

Serves 4

Ingredients

For the pheasant Wellington
2 shallots, peeled
5–6 ounces (about 150 g) button mushrooms
2 cloves garlic
2 tablespoons (30 ml) olive oil
1 pound (450 g) puff pastry sheets
4 skinless pheasant breast filets
1 large egg, beaten with 1 tablespoon (15 ml) water

For the red cabbage
½ red cabbage, shredded and cored
1¼ cups (300 ml) red wine
½ cup packed (100 g) soft brown sugar

Chef's Tips

For the mustard mash, boil 4 large, peeled, chopped red potatoes until tender. Drain, then mash with ½ stick (60 g) butter and 1 tablespoon (15 ml) whole grain mustard. Season with salt, to taste.

Method

1 To make the Wellington, blend together the shallots, mushrooms, and garlic in a food processor until coarsely blended. Heat the oil in a frying pan, then add the vegetable mix and cook gently, stirring regularly, until softened. Cool.

2 Preheat oven to 350°F (180°C). Cut the pastry into four portions. Spoon over the mushroom mix, then top with one pheasant breast each. Fold over pastry and brush edge with egg wash to create sealed parcels. Brush remaining egg over the pastry to glaze. Transfer to a baking sheet. Bake for 15–20 minutes, or until pastry is golden.

3 Meanwhile, simmer the shredded cabbage in the wine and sugar for 20 minutes, or until the liquid evaporates and the cabbage is soft and tender. If the cabbage dries out before it is cooked, simply add extra water to the pan. Serve with the pheasant Wellington and mustard mashed potatoes, see Chef's Tips.

Roast duck with celeriac & burned apple

Burning apples may sound strange, but it brings out a wonderfully sweet acidity that complements roasted duck and earthy celeriac. Use the roasting juices to make a tasty gravy.

Serves 4

Ingredients

1 duck, about 6½ pounds (3 kg)
salt and black pepper
1 teaspoon (5 g) freshly ground caraway seeds
1 firm dessert apple, peeled, cored, and chopped
1 lemon, chopped
1 onion, chopped
1 clove garlic, crushed
1 sprig thyme

For the celeriac & burned apple
1 celery root (celeriac), peeled and cubed
2 tablespoons (30 ml) olive oil
salt
3 sprigs thyme
1 clove garlic, peeled and crushed
3 Granny Smith apples

Method

1. Preheat oven to 350°F (180°C). Pierce the skin of the duck all over with a skewer. Season with salt, pepper, and caraway. Stuff the cavity with the fruit, vegetables, and thyme. Sit the duck breast-side down on a wire rack over a deep baking pan.

2. Roast duck for 40 minutes, then turn it breast-side up. Pour off the fat and reserve. Reduce oven to 320°F (160°C). Roast duck for 1½–2 hours, until core temperature is 150°F (65°C). Rest duck in a warm place for 30–40 minutes. Carve off the two breasts and slice them in half lengthwise.

3. Place the celery root (celeriac) in a baking pan with the oil, salt, thyme, and garlic. Cover and bake with the duck for 60 minutes, or until tender.

4. For the burned apple, peel, core and cut each apple into thick slices. Heat a nonstick frying pan, then add the apples and char for 5 minutes to brown. Cool in the pan without moving the apple. Remove apple with a spatula, and place burned-side up in a baking pan with the celeriac. Reheat briefly in the oven before serving with the duck.

Knives

Sharp knives are the key to cutting and carving. If your knives are really dull, you'll need to sharpen them on a whetstone to make a new sharp edge. To keep them sharp, use a steel once a week or so. Holding your steel in your left hand and your knife in your right, tilt the knife at a 20° angle to the steel and draw it down the steel moving your elbow back at the same time. This movement sharpens the whole length of the knife. Repeat a few times until you get a nice, sharp edge. Never store your knives in a kitchen drawer where they can become dulled by bumping into other things (and where they could cut you); use a knife block or a magnetic knife strip.

Is it cooked yet?

Using an instant-read meat thermometer is the most accurate way to tell if meat is cooked. When the cooking time is up, remove the food from the heat and stick the thermometer into the thickest part of the meat, avoiding any fat or bones, then wait 20 seconds for the temperature to register on the screen. If the meat is not yet at the desired temperature, check it again 10 minutes later. After use, wash the needle of the thermometer in warm soapy water.

INTERNAL TEMPERATURES

The USDA recommends cooking all meat thoroughly to avoid food-borne illness. For beef, veal, pork, and lamb steaks, chops, and roasts, the approved temperature is 145°F (63°C); for all ground meat, it's 165°F (75°C). The recommended temperature for cooking poultry is set at 165°F (75°C). Some people prefer their meat more or less well-done and the chart below provides a guide.

BEEF

rare: 120–125°F (49–51.5°C)

medium-rare: 130–135°F (54.5–57°C)

medium: 140–150°F (60–65.6°C)

medium-well done: 155°F (68.3°C)

well-done: 160–170°F (71–74°C)

VEAL

medium: 145–155°F (63–68.3°C)

PORK

medium: 145°F (63°C)

well-done: 180–185°F (82–85°C)

LAMB

medium-rare: 130–135°F (54.5–57°C)

medium: 140–150°F (60–65.6°C)

medium-well-done: 155–160°F (68.3– 71°C)

well-done: 165–185°F (74–85°C)

BUFFALO AND VENISON

rare: 120–130°F (49–54.5°C)

medium-rare: 130–140°F (54.5–60°C)

As they are low in fat and can easily dry out, buffalo and venison should not be cooked beyond medium-rare. Take care not to overcook.

CHICKEN AND TURKEY

breast: 165°F (75°C)

legs: 170°F (76.7°C)

Chicken breast meat can dry out, so it does not need to be cooked to as high a temperature as the legs. If cooking with stuffing inside the bird, the stuffing should reach at least 135°F (57°C).

DUCK

medium-rare: 155–165°F (68.3–75°C)

medium: 165°F (75°C)

The chefs

Ruby & White supply the most highly regarded restaurants across England's West Country. The skilled chefs who lead these kitchens have shared some of their most popular recipes for this book.

NICK ARMITAGE

Nick is Chef Patron of The Town House restaurant, winner of the Best British and Best Sunday Lunch at the Bristol Good Food Awards. A Somerset native, Nick trained at Leith's School of Food and Wine, London, before moving to New Zealand where he was a part-owner of The Reef in Dunedin. Now back in England, Nick can be found making piles of perfectly puffed Yorkshire puddings every Sunday. **www.thetownhousebristol.co.uk**

DAVID BROWN

David is Sous Chef at the Manor House Hotel, Moreton-in-Marsh, Gloucestershire. He had previously worked at the Gleneagles Hotel and Restaurant Sat Bains. David uses Asian influences to redesign British dishes. His food philosophy is simple … "nature provides us with some amazing produce so I try to showcase the food with slight manipulations." **www. cotswold-inns-hotels.co.uk/property/the_manor_ house_hotel**

ALISON GOLDEN

Alison is Head Chef at the Circus Cafe and Restaurant, Bath. Ali has been running her own restaurants for 30 years, and believes in sourcing fantastic ingredients and cooking them simply with respect for the essential flavors. Her three food guiding lights have been Elizabeth David, Jane Grigson, and Joyce Molyneux. Her dishes have been tried and tested over the years. **www.thecircuscafeandrestaurant.co.uk**

JOHN HOGAN

Executive Chef at Keefer's Restaurant, Chicago, John has years of experience at Chicago's finest restaurants, and dozens of awards. His love of food dates back to his childhood when he watched his mother cooking special meals for his father. John is known for innovative haute cuisine and simple fare with his signature twists. For Keefer's, Hogan developed a new menu of Chicago-style steak and seafood dishes. **www.keefersrestaurant.com**

DAVE KELLY

Dave may not be a chef, but his years in the butcher's trade mean that he knows meat. For this book, Dave has written down some of the recipes that he cooks for family and friends. Simple and satisfying, they're some of his favorites, and he hopes they'll be yours too. **www.rubyandwhite.com**

LUIGI LINO

Luigi is Head Chef at Martini Ristorante, voted the Best Italian Restaurant in Bath, England, at the UK's Good Food Awards. Luigi was born in Naples, but grew up in Rome. He studied cooking in Rieti before working in the north of Italy and in Bermuda. Luigi loves cooking meat, particularly beef, and his passion for Italian food comes from his mother and grandmother. **www.martinirestaurant.co.uk**

SAM MOODY

Sam is Executive Chef at Bath Priory and was awarded his first Michelin star in 2012. Sam has a reputation for exciting flavors and the delivery of a culinary experience that delights time and again. He develops new menus with fresh, local ingredients, and values strong supplier relationships—his locally grown salads are sown to his own precise requirements—the link between food and cuisine does not come closer than this. **www.thebathpriory.co.uk**

NICK ORR

Head Chef at the Manor House Hotel, Moreton-in-Marsh, Gloucestershire for four years, Nick first studied catering in Stratford-Upon-Avon. He creates classic British-continental cuisine with his own modern twist. Off-duty, Nick loves comfort food: coq au vin, casseroles, and stews. At the restaurant, he creates and teaches complex modern dishes to his dedicated team. **www.costwold-inns-hotels.co.uk/property/the_manor_house_hotel**.

ROSS SHAW

Ross is Head Chef at award-winning Gascoyne Place, Bath. The restaurant specializes in modern British cuisine, using fresh, seasonal, locally sourced produce. Ross is a fan of modern catering techniques, utilizing a water bath (sous vide) to retain the essential flavors and textures of ingredients. He loves working with game, particularly venison. **www.gascoyneplace.co.uk**

GENEVIEVE TAYLOR

Genevieve is a food writer and customer at Ruby & White butchers. She loves simple, colorful, gorgeous food—food that looks as good as it tastes. She cooks with seasonal ingredients and much of her inspiration comes from a lifelong love of the outdoors. Gen is the author of *A Good Egg* and three other cookery books. She writes a monthly cookery column for *Crumbs* magazine and lives in Bristol, England. **www.genevievetaylor.co.uk**

Index

CHEF CREDITS

Nick Armitage of The Town House
Classic Sausages with Bubble & Squeak, p.98
Duck and Sarladaise Potatoes, p.205
English Roast Beef & Yorkshire Pudding, p.70
Hangar Steak with Peppercorn Sauce, p.32
Lamb Two Ways, p.120
Rabbit & Ham Terrine, p.148
Roast Beef on Hot Dripping Toast, p.72
Roast Pork Shoulder with Rhubarb & Ginger Sauce, p.102
Sticky Chicken Wings, p.172
Warm Pigeon Salad, p.200
Wild Boar, Guinea Fowl & Apple Meatballs, p.156

David Brown and Nick Orr of The Manor House Hotel
Coq au Vin, p.190
Herbed Pork Patties, p.96
Lamb Broth with Croquettes, p.118
Oxtail with Tagliatelle, p.60
Pheasant Wellington & Red Cabbage, p.211
Sticky Spareribs, p.91

Alison Golden of The Circus Cafe and Restaurant
Lamb Tagine, p.126
Oxtail Crumble, p.61
Rabbit Pie, p.146
Seared Calf's Liver with Anchovies, p.49
Sweetbread Fritters on Banana & Apple purée, p.48

John Hogan of Keefer's
Braised Oxtail with Marrow-Whipped Potato, p.56
Braised Pork Cheek with Green Lentils & Root Vegetables, p.88
Breast of Duck with Duck Cake & Compote, p.202
Brioche-Crusted Sweetbreads with Leeks & Truffle Sauce, p.46
Chili-Rubbed Smoked Pork Belly with Chipotle Sauce, p.100
Chilled Vietnamese Braised Beef Shank, p.65
Lamb Sausage with Marble Potato Salad, p.122
Sirloin with Sauce Verte, p.34
Leg of Chicken "Crépinette", p.174
Pretzel-Crusted Pork Chop Viennoise, p.84
Rabbit Rillettes Niçoise, p.144
Roast Chuck Eye with Potato-Mushroom Gratin, p.68
Roast Rack of Lamb with Summer Vegetable Ragout, p.130
Roast Venison with Oatcake & Huckleberry Sauce, p.150
Veal Medallions with Morels, p.36

Dave Kelly of Ruby & White
Dave's Best Beef Burger, p.41
Beef Shank with Root Vegetables , p.64
Boned and Rolled Chicken, p.181
Butter-Roasted Chicken Supremes, p.171
Chicken Pie, p.180
Classic Beef Stock, p.17
Cottage Pie, p.15

Golden Crumbed Chicken with Cheese, p.177
Honey-glazed Pig's Tails, p.90
Pefect Porterhouse Steak, p.30
Perfect Sirloin Steak, p.31
Pork Belly Braised in Cider, p.97
Slow-Roasted Lamb Shoulder, p.127
Stuffed Pork Loin Chops, p.104
Sweet Cheat's Chili, p.63

Luigi Lino
Agnello Scottadito, p.124
Bistecca alla Fiorentina, p.35
Boar Stew with Blueberries, p.161
Classic Neapolitan Ragu with Involtini, p.86
Deviled Chicken Livers, p.187
Ossobuco, p.58
Pork Chops with Apple & Cider, p.85
Roasted Chorizo-Stuffed Chicken, p.178
Saltimbocca alla Romana, p.37

Sam Moody
Anchovy Lamb with Cauliflower "Couscous", p.132
Boiled Ham Hock & Pease Pudding, p.98
Braised Beef Cheeks, p.62
Roast Duck with Celeriac & Burned Apple, p.212
Salt Beef & Red Onion Marmalade Sandwich, p.43

Ross Shaw
Venison Burgers with Root Remoulade, p.154

PHOTO CREDITS

Ian Armitage: 218,TL. Getty Images: 2 Louise Lister/StockFood Creative; 38 Brian Leatart/Stockfood Creative. Jason Ingram: 219, BR. Keefer's: 7, 10. Sean Malyon: 218, BL. Mary Evans Picture Library: Cover, 5, 17. Photocuisine: 129 Mallet. Superstock: 19 Tips Images; 67 Marka; 77 age footstock; 93 Food and Drink; 135 Tombini Marie-Laure/Oreida Eurl; 153 Food and Drink; 159 Stockbroker; 183 Marka; 185,189, 207, 209 Food and Drink. Shutterstock: 45 Fotografiche; 47 bogumil; 51 violeta pasat; 55 Anna Kurzaeva; 57 Darkkong, TL; 69 BW Folsom, TL; Juan Nel, BR; 89 Andreja Doniko; 101 B. and E.

Dudzinscy, TL; surabhi25, BR; 108 David Young; 123 Jultud, TL; Diana Taliun, BR; 131 Claudio Zaccherini,TL; Marco Mayer, BR; 139 JKlingebiel; 141 Critterbiz; 151 Jiang Hongyan, TL; IngridHS, BR; 152 Francesco83; 155 Kuttelvaserova Stuchelova; 165 Naffarts; 168 James BO Insogna;175 alekleks, TL; ben44,BR; 195 Ansis Klucis,196 Paul Cowan ML; 197 Tobik. TR; papa 1266, BR; 203 HLPhoto, TL, Ann Baldwin, BR; 214-215 Thomas M Perkins; 216 Lilyana Vynogradova.

All other pictures by Mark Winwood.

With special thanks to Vanessa Addam, Jake Dickson at Dickson's Farm Stand (www.dicksonfarmstand.com), and Dan Marino at Jackson Hole Buffalo Meat (www.jhbuffalomeat.com).